View Chapter Summary Videos from T. Yomi Obidi:

http://books.sae.org/r-392/

Theory and Applications of Aerodynamics for Ground Vehicles

Other SAE books of interest:

Aerodynamics of Road Vehicles, Fourth Edition
By Wolf-Heinrich Hucho
(Product Code: R-177)

Fundamentals of Vehicle Dynamics
By Thomas D. Gillespie
(Product Code: R-114)

Race Car Vehicle Dynamics
By William F. Milliken and Douglas L. Milliken
(Product Code: R-146)

Theory and Applications of Aerodynamics for Ground Vehicles

By T. Yomi Obidi

SAE INTERNATIONAL

Warrendale, Pennsylvania
USA

400 Commonwealth Drive
Warrendale, PA 15096-0001 USA

E-mail: CustomerService@sae.org
Phone: 877-606-7323 (inside USA and Canada)
724-776-4970 (outside USA)
Fax: 724-776-0790

ISBN 978-0-7680-2111-0
SAE Order Number R-392
DOI 10.4271/ R-392

Library of Congress Cataloging-in-Publication Data
Obidi, T. Yomi, 1955-
Theory and applications of aerodynamics for ground vehicles / by T. Yomi Obidi.
pages cm
Includes bibliographical references and index.
ISBN 978-0-7680-2111-0
1. Automobiles--Aerodynamics. 2. Trucks--Aerodynamics. I. Title.
TL245.O25 2014
629.2'31--dc23
2013044452

Cover photo courtesy of Jaguar Land Rover/Exa Corporation. Used with permission.

Dedication

To
Ọnayin-yi Oshomoshi Olodumare

Contents

Preface

The focus of this book is on cars and trucks, these being the most common ground vehicles in the speed range in which the study of ground vehicle aerodynamics is beneficial. With an understanding of ground vehicle aerodynamics, one is equipped with a basis for the design of vehicles with better fuel efficiency, with improved vehicle performance, and with increased passenger comfort.

Naturally, many of the materials presented herein can be found in other books on ground vehicle aerodynamics or even in aerodynamic books in general. There are, however, some topics that are either skipped or treated rather lightly, the same topics that I consider deserving of a more rigorous treatment than has been found in the few books that address them. One such topic is *car-truck interaction*, when one vehicle (usually the smaller one) is overtaking the other. There is a direct and instant benefit in terms of safety on the highway from understanding the forces at play when one vehicle passes the other in the same direction and sense.

Also, most of the books on ground vehicle aerodynamics appear to target the practicing engineer or aerodynamicist, leaving aspiring students wondering more than implementing at the time that they are still in school. This book aims to introduce the student to ground vehicle aerodynamics and to methodically guide the student through the school days to the practicing level, at which stage there are a wider range of books to consult.

The material in this book has been previously presented, in part, at seminars and in professional development sessions.

T.Y.O.

Acknowledgments

The works of pioneers often appear to be of little significance at the time that the work is done. The works of pioneers are often pieces, formed from thoughts, looking for other pieces to form a body of thought, if complimentary, or a worthy provocateur, if different, yet challenging. Together they form the foundation for a discipline. This book has enjoyed the benefits of the foundation laid by the pioneers of aerodynamics in general, and of the aerodynamics of ground vehicles in particular. Some of those pioneers are well known, while others are unsung, and unpublished, yet valuable.

It is my hope that this book bears testimony to the diligence of the forerunners in the field of ground vehicle aerodynamics, and that it provides some light for those yet to come, as well as for others in the field.

I am profusely thankful to John Fischer, who volunteered both time and resources to facilitate the completion of this book. I am very thankful to Martha Swiss of SAE International (SAE) for her unusual patience through what almost became a perpetual shift of manuscript due dates. Thanks to Heather Slater and to Bill Schall, both of SAE, for critically reviewing the manuscript and ensuring proper drawing format, respectively. My dry days were lightened with prayers, encouragement, and assistance from family and friends—Papa, Mama, Funlola, Shola, Tosin, Teresa, "the kids" (from Tunmishe through Onoye), Uncle Dele, Aunty Marcy, Aunty Glendora, Uncle Bill (Babake), Dr. Joshua Owoeye, Aunty Margaret Omodara, Chief Adesola Awolola, and many more. May heaven smile upon you in your times of need.

T.Y.O.

Introduction

The necessity for improved vehicle fuel economy is a major motivator in the attempt to better understand ground vehicle aerodynamics. Drag, lift, and stability are three concepts that constitute the cornerstone in the study of ground vehicle aerodynamics. The consideration of stability typically influences the design of control surfaces, a common feature in race cars. Though lift is a desirable occurrence in flight, groundedness, an inverse of lift, is the goal in ground vehicle design. This negative lift is often referred to as *downforce* in ground vehicle aerodynamics.

Perhaps the single largest aerodynamic consideration in the design of ground vehicles is drag. A major goal of the effort to better understand ground vehicle aerodynamics is the optimum reduction of aerodynamic drag. The benefits of a successful drag reduction are instant. They include better fuel efficiency, improved vehicle performance, and increased passenger comfort.

What really is *drag*? To answer this question, perhaps we should first treat the question, *What really is aerodynamics?*, and, by extension, address the concept of *ground vehicle aerodynamics*. Aerodynamics is simply the study of the forces involved in the movement of an object through the air. Various objects—airplanes, cars, trains, footballs, cricket balls, tennis balls, baseballs, feathers, even *agbada* (Fig. 1.1)—interact with and are therefore affected by the dynamics of the surrounding air. Because of the general implication of the term *vehicle*, vehicle aerodynamics narrows our consideration to transportation. Ground vehicle narrows it even further to vehicles that make contact with the ground throughout their movement. So ground vehicle aerodynamics is the aerodynamic study of cars, trains, trucks, trailers, motorcycles, carts, bicycles, and, lest we forget, human beings—walking, running, or crawling.

Air molecules bombard everything that they encounter and flow against or past it. When the air molecules are not constrained, the flow is generally termed incompressible, and when this incompressible flow travels past a cooperative object such as an airfoil, we have a streamline of the flow. Streamlines are typically laminar flow. As flow speed increases rapidly or as obstruction to flow increases in the form of friction or in the form of a blunt obstacle, the tendency for turbulence increases. The streamlines soon give way to micro flow reversals and the build-up of the shear layer at the incipience of turbulence.

Be it laminar or turbulent flow, drag is generated as a reaction to flow, and all objects struggle to overcome this drag, which is either in the form of pressure or friction. The pressure form of drag is dominant in flow against blunt objects such as a flat plate or a bluff body positioned head-on against the flow. Drag in the form of friction is the most common form of drag experienced by moving bodies, the latter having evolved to minimize pressure drag.

Although the airfoil is considered the most efficient form of design for drag reduction, ironically, protrusions from an almost-airfoil design actually assist in reducing the drag on the body, especially at high speed. Just like the spinning stitches of a baseball assist the ball in its movement through the air by breaking up the air, the extruded parts of a vehicle—the door handle and mirror, for example—also help to break up the air and reduce air resistance. Drag reduction, which is the primary aerodynamic design goal of a vehicle, is therefore not merely a straightforward horizontal tear-drop design exercise but a comprehensive approach in which particular attention is paid to parts of the vehicle at different locations such that the collective is well integrated into the aerodynamic *big picture*.

Written for senior-level undergraduate students and graduate students, this book serves well the practicing engineer, the aerodynamicist, the vehicle designer, and the vehicle rendering artist. It consists of 13 chapters, some of which can be skipped without loss of continuity. Chapter 1 is primarily aerodynamic review. Naturally, it starts with the treatment of the subject of drag. It goes on to describe the triggers and the consequences of drag in general and in ground vehicles in particular. Chapter 1 may be skipped if the reader already has an aerodynamic background. Chapter 2 discusses the effects of noise and vehicle soiling on ground vehicles, particularly on cars. Passenger comfort is of a greater interest (than aerodynamics) in this chapter. Chapters 3 discuses wind tunnel testing as well as track and road testing, both experimental methods for vehicle testing. Wind tunnel types and functions are introduced. Inherent and potential errors in wind tunnel testing are addressed, and the correction methods are presented. Design methods of rigs for road tests and data collection methods for road tests are presented. Chapter 4 is an introduction to numerical methods. A case is made for CFD over wind tunnel and road test methods in particular settings. Various types of computational models are listed with their respective benefits and deficiencies. The sequence of numerical solution from model building through boundary layer prescription to solution and result interpretation is treated. In chapter 5, topics of vehicle stability and control and vehicle performance are treated. Primary control surfaces such as spoilers, winglets, and endplates are presented. Secondary control surfaces and media such as vehicle underbody and cross wind are treated. A detailed treatment of truck-car interaction is presented based on extensive studies by the author. Chapter 6 brings together background knowledge from earlier chapters into the design of an aerodynamically sound car design, with the car broken into front, mid-, and rear assemblies. The hood, windscreen, nose, and fascia assembly make up the front section. The roof and the cabin-sustaining posts (A, B, C, and sometimes D) make up the mid-section, and the trunk and rear bumper constitute the rear section. Chapter 7 may be considered as an introduction to the aerodynamics of large vehicles. From light trucks to trailers and buses, aerodynamic factors in the vehicle design are presented. A detailed aerodynamic analysis of side skirts—a recent add-on to large trucks—is presented. The topic of car-truck interaction begun in chapter 5 is given further treatment due to a crucial safety issue that the author feels is at stake in highway car-truck common *environment*.

If chapter 7 is an introduction to large vehicle aerodynamics, chapter 8 is a continuation of large vehicle aerodynamics, with a focus on trains. Low-speed, high-speed, and very high speed passenger trains are discussed. Vibration is a current problem in high-speed trains as they travel through tunnels. The forces and moments that trigger such vibration are presented as well as the design features that have the potential to ameliorate them. Chapters 9 and 10 are at opposite ends of the aerodynamic spectrum. Severe service and off-road vehicles are treated in chapter 9, while chapter 10 looks at race cars. Even though aerodynamics is not what comes to mind in the design of severe service and off-road vehicles, there are inexpensive modifications that can be made with a resulting improvement in fuel efficiency. An example of such improvements is found in the aerodynamic profiling of the underbody of severe service vehicles. Race cars and sports cars are the epitome of aerodynamic ground vehicles. It is recommended to review chapter 5 before embarking on chapter 10. Motorcycle aerodynamics is introduced in chapter 11. The same features that have been discussed in earlier chapters are here applied to motorcycles. The seriousness of the safety issue in motorcycle-truck interaction is underscored by the attention given to it in the treatment that follows up on chapters 5 and 7. Chapter 12 treats the subject of internal aerodynamics and cooling system flow in ground vehicles. Although some aspects of this chapter are treated in chapters 5 and 6, the chapter, like chapter 2, can be studied independently. Chapter 13 is an open-ended chapter. Rather than the title, *Concept Ground Vehicles*, it may very well be titled, *The Quest Continues*.

Chapter 1
Drag

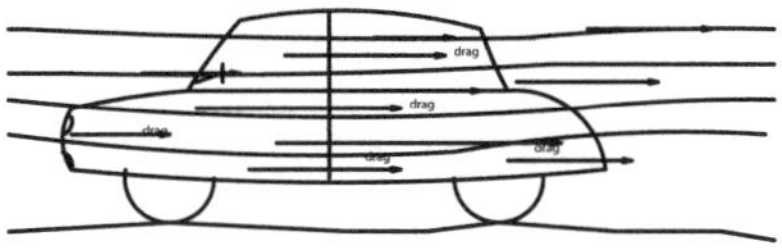

The primary goals of the study of ground vehicle aerodynamics are the reduction of aerodynamic drag and the improvement of vehicle safety. The safety of a vehicle is fundamental to its value as a tool of transportation. Similarly, the reduction in the drag of a vehicle provides tangible benefits that add to the value of the vehicle, including better fuel efficiency, improved vehicle performance, and increased passenger comfort.

1.1 General Concept of Drag

In general, a drag is a force that works against the intended or desirable direction of the entity that experiences the drag. It manifests in the form of pressure or friction. All objects struggle to overcome drag (Fig. 1.1). However, the greater the struggle, the greater is the drag. A car's struggle to overcome drag is in the form of increase in speed. But the higher the speed, the greater is the drag. How then does the car get the benefit of speed over drag? For one, the thrust increases at a greater rate than the drag. The other benefits are found in the design of the car and in the material of the body. An aerodynamic design will incur a drag that is much less than a bluff body design (Fig. 1.2(a), (b)), other factors being the same. Also, a rough surface causes a different drag pattern than a smooth surface, with a smooth surface initially showing lower drag until, after some critical speed, the drag on the rough surface suddenly drops to a value lower than that of the smooth counterpart and continues in that fashion as both drag forms increase with further increase in speed [1-1, 1-2].

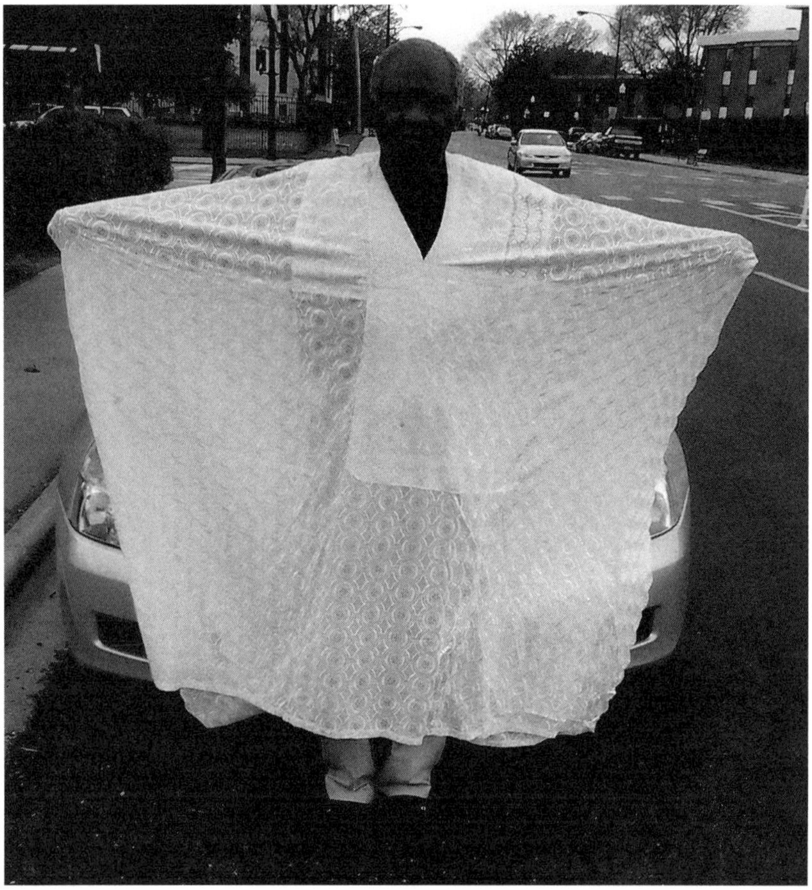

Fig. 1.1 Aerodynamic drag

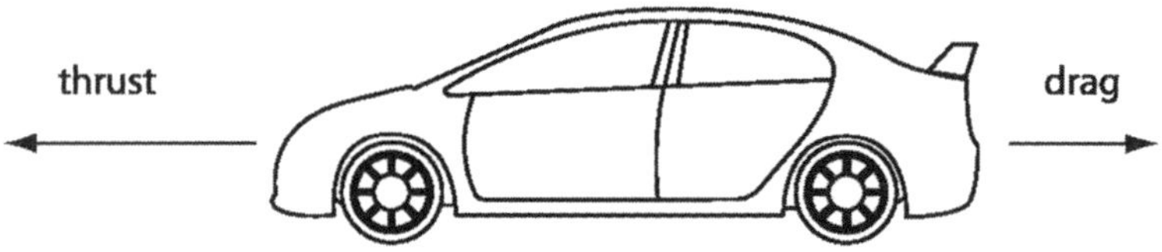

Fig. 1.2(a) Forces on a car with aerodynamic features

Fig. 1.2(b) Forces on a bluff-body car

The increase in drag with speed is not indefinite. As the speed increases, at some point the drag begins to drop. Thereafter, it starts to rise again with a continuously increasing speed, but on a different curve from the first (Fig. 1.3). The first curve depicts the drag during laminar flow, whereas the new curve formed after the region of uncertainty is the turbulent flow curve. The flow has transitioned from laminar to turbulent flow.

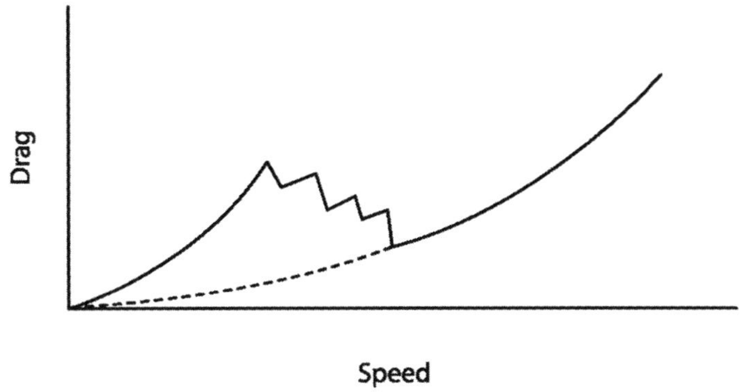

Fig. 1.3 Variation of drag with speed

In laminar flows, the fluid particles move downstream in smooth and regular trajectories with almost no mixing between different layers of the fluid. The flow is steady: each fluid particle passes through a given point in space, arriving there (at that point) with exactly the same speed and flow direction as did all other particles before it. (More will be said about this in Section 1.6.) In turbulent flows, however, mixing and interchange of mass and momentum between different streamlines results ultimately in a higher drag.

It would seem then that if one were to streamline an object well enough, one could accomplish zero drag. But that is not the case, even at a low speed. Therefore, there have to be other factors contributing to drag incurrence. Airfoil theory has largely helped to facilitate the understanding of aerodynamic flow. We will therefore use an airfoil to explain the forms of drag that an object experiences. A detailed discussion of the airfoil will be presented in the section on airfoil theory. It suffices for now to understand the basic shape of the airfoil as a horizontal teardrop (Fig 1.4(a)).

Air particles at atmospheric pressure flow toward the airfoil, making a first contact with the leading edge of the airfoil at the nose (Fig. 1.4(b)). Here, the flow is momentarily stagnant as all the dynamic energy of the flow is converted to pressure. The nose is appropriately referred to as the stagnation point. The flow is accelerated downstream between the nose and shoulder (i.e., the point of maximum thickness of the airfoil) as the pressure drops. By the time a particle reaches the shoulder, the pressure is reduced to its minimum value, and the

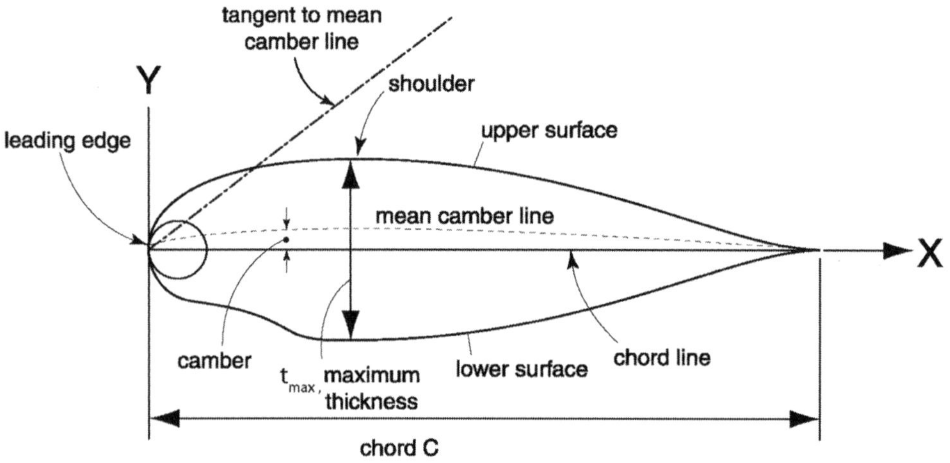

Fig. 1.4(a) Airfoil geometry

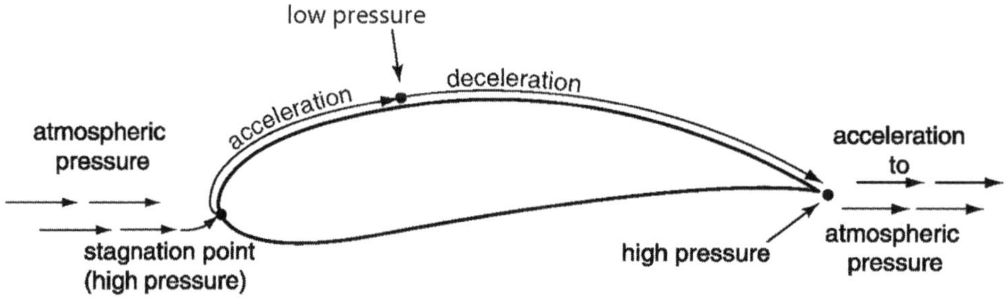

Fig. 1.4(b) Simplified flow pattern on an airfoil

momentum becomes much less than that of the opposing viscous force it experiences. Therefore, as the fluid moves from the shoulder to the tail, a boundary layer particle has a reduced momentum and opposes the viscous and decelerating forces caused by the increasing pressure. With continuously increased pressure, the boundary layer particles may come to a standstill, a condition referred to as a stall.

A streamlined shape ideally eliminates the occurrence of boundary layer separation. When a streamlined airfoil is used, the rate of pressure rise from the shoulder to the tail is so moderate that the boundary layer continues all the way to the tail without the airfoil experiencing stall. In this case, the flow outside the boundary layer is more or less the same as a non-viscous flow; thus there is no pressure drag.

The boundary layer is a region of retarded flow. Outside the boundary layer, the flow is unencumbered by viscosity. The greater the speed, or the smaller the viscosity, the less rapidly the boundary layer grows. The Reynolds number is perhaps the most prominent of the terms that provide a relationship between the determining parameters of the boundary layer of a flow.

Reynolds number, Re, is defined by

$$\mathrm{Re} = \frac{\rho \, \mathrm{v} \, \mathrm{d}}{\mu} \text{ or } \mathrm{Re} = \frac{\mathrm{v} \, \mathrm{d}}{\upsilon}$$

where

ρ is the density of the fluid;

v is the speed at which the fluid particles move;

d is a consistent length (usually the chord of a selected airfoil);

μ is the fluid dynamic viscosity (N s/m^2 or kg/ms); and

υ is the kinematic viscosity (m^2/sec).

Because Reynolds number is a convenient indicator of the probability of a flow remaining laminar or turning turbulent, we can, for example, increase the Reynolds number to a desirable value by speeding up the flow or by smoothening the surface that the flow travels over, thereby reducing the viscosity.

1.2 High Reynolds Number Flow versus Low Reynolds Number Flow

Stokes flow is a low Reynolds number flow. From Stokes law, the drag on an object is a function of the speed, the viscosity, and the size.

$$\text{Drag} \sim \mathrm{v} \, \mu \, \mathrm{d}$$

The viscosity of air (or a gas in general, for that matter) decreases with increasing temperature. A large change in pressure at a constant temperature does not cause a significant change in the viscosity of a gas. There are various forms of expressing viscosity variation, but they all take the general form

$$\mu(T) = \mu(T_o) - BT^n$$

$\mu(T_o)$, B, and n being constants, and T_o a reference temperature.

At high Reynolds numbers (Re >> 1), the drag is a function of the density, speed, size, and (to a smaller extent) the viscosity. In flows with high Reynolds numbers, the drag is proportional to the square of the speed, and the power (which is proportional to both drag and speed) is proportional to the cube of the speed. The mention of power is significant here because of the direct correlation between power and fuel consumption. Fuel consumption is therefore proportional to the cube of the speed. Table 1.1 contrasts the drag at high Reynolds number with the low Reynolds number case.

Table 1.1 Low Re versus High Re Drag Dependence

Drag at Low Re	Drag at High Re
Proportional to speed	Proportional to $(\text{speed})^2$
Proportional to size	Proportional to $(\text{size})^2$
Proportional to viscosity	Proportional to density
(Density not significant)	(Viscosity not significant)

Using the principle of dynamic similarity, which states that if the Reynolds numbers are the same then the drag coefficients are the same, we can construct meaningful model tests and correctly apply the results to the full-scale object.

1.3 General Airfoil Flow Behavior

As the Reynolds number increases, the boundary layer grows less rapidly. However, the viscous boundary layer never disappears. This is because the velocity at the wall remains zero, a condition referred to as the *no slip condition*. With increasing Reynolds number, a point is reached at which there is flow separation. The boundary layer separation at high Reynolds numbers is a result of pressure distribution in which the pressure drops from the nose to the shoulder of the airfoil, then rises from the shoulder to the tail.

In contrast, there is almost no separation at low Reynolds number flows. In this case, the pressure drops from the nose to the shoulder, and continues the drop all the way to the tail (see later Fig. 1.8(b)). The falling pressure pushes the low momentum fluid near the surface, so that there is no backflow, hence, no flow separation.

At high Reynolds number, a streamlined body has negligible pressure drag because the average pressure is about the same everywhere (Fig. 1.5). With non-streamlined round objects—sphere, rod—the rate of pressure rise from the shoulder to the tail is so high that the boundary layer becomes stalled quickly

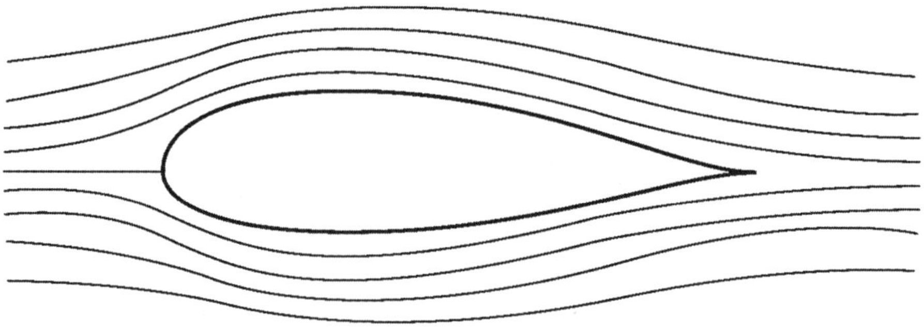

Fig. 1.5 Boundary layer for a practically non-viscous streamline flow

right by the shoulder (diameter of the round object). A broad wake of eddies forms downstream of the separation (Fig. 1.6). This wake is much wider than the viscous wake of the streamlined shape. Also, because of the blockage of the mean flow caused by the relatively broad space occupied by the wake, the flow pattern is changed. The pressure on the aft of the body is no longer equal to that on the front. It is higher than on the shoulder but lower than on the front. The net pressure drag connected with the flow separation is usually several times greater than the viscous drag due to friction at the surface of the body, and it is the dominant drag of non-streamlined bodies.

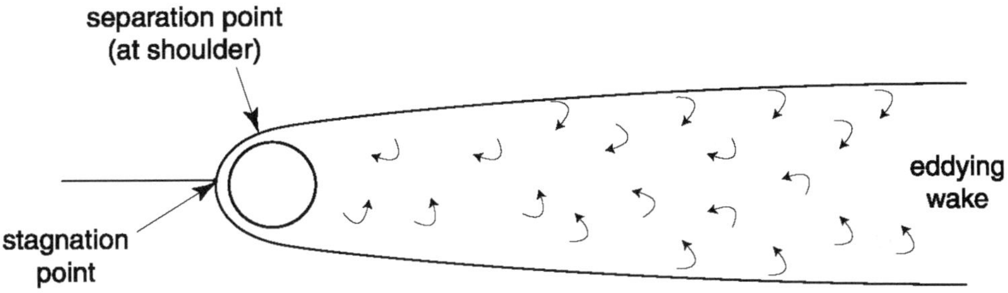

Fig. 1.6 Flow over a sphere

The same result can be expected in a well streamlined body by adversely increasing the angle of attack to increase lift (Fig. 1.7(a), (b)). At high Reynolds numbers, there is separation due to pressure distribution produced by the main body of the non-viscous flow as the flow system seeks to create pressure-inertia dynamic balance outside the boundary layer (Fig. 1.8(a)). At low Reynolds numbers, however, there is no separation. This is because the inertial forces are negligible as viscosity dominates, and the flow is governed by the static balance of pressure-viscous forces, resulting in a different form of pressure distribution on the airfoil (Fig. 1.8(b)). The constantly falling pressure pushes along the low momentum fluid near the surface so that there is no backflow and hence no separation. Streamlining does not change the pressure drag to a significant extent at low Reynolds numbers; rather, it increases the viscous drag at the surface of the body because of the increased surface area over which the frictional stresses act. Because of this increase in the wetted area, the total drag is actually increased by streamlining at low Reynolds numbers.

At high Reynolds numbers (which usually result in turbulent flow), most of the drag is pressure drag. A turbulent boundary layer has a great deal of mixing in it. The mixing generates a momentum exchange by means of which the low-speed boundary layer fluid can borrow momentum from the high-speed fluid outside the boundary layer. Due to turbulence, the area of low pressure on the downstream side is narrower than in the laminar flow. In addition, the pressure on the downstream side in the laminar flow is lower than in turbulent flow (Fig. 1.8(b)).

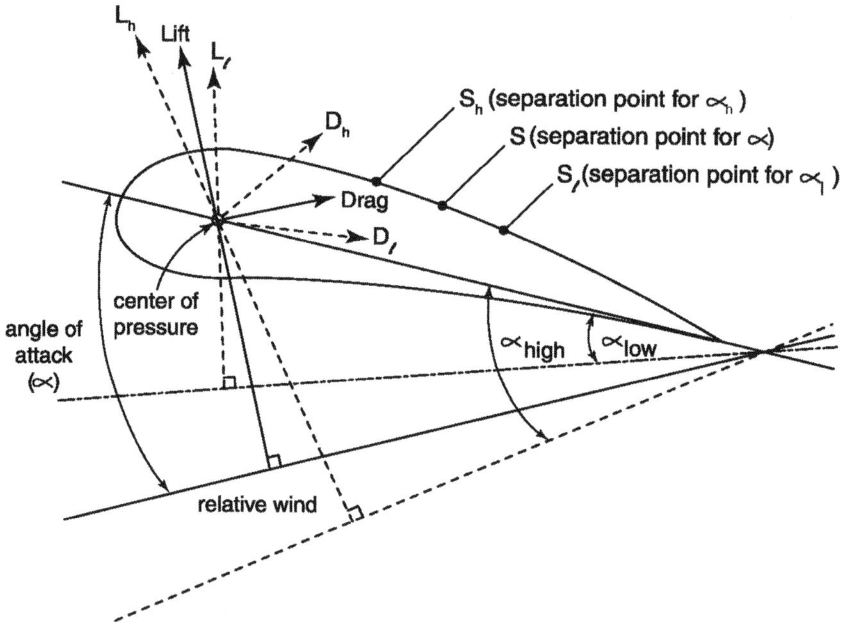

Fig. 1.7(a) Effect of varying the angle of attack (α) on lift (L) and on flow separation

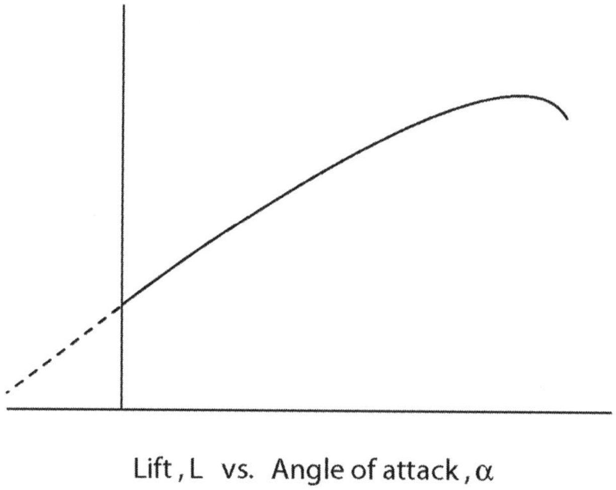

Fig. 1.7(b) Lift, L, versus angle of attack, α

The generalized concept about the flow over an airfoil seems to be in question following the two instances just reviewed—one with increased drag in a laminar flow and the other with delayed separation in a turbulent flow—until the premise of each result is carefully viewed. At low Re, the flow is attached longer on the airfoil. It therefore has a greater wetted area subjected to viscous

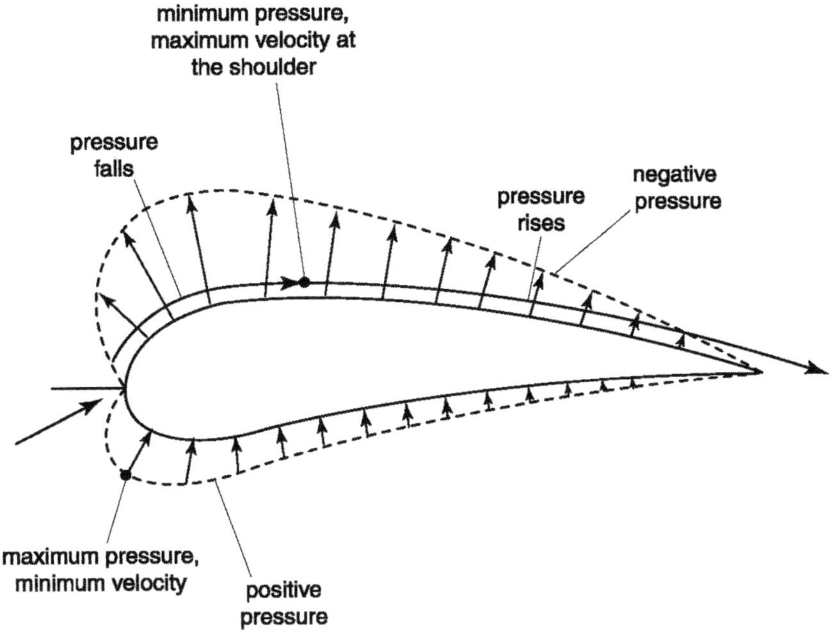

Fig. 1.8(a) High Re flow

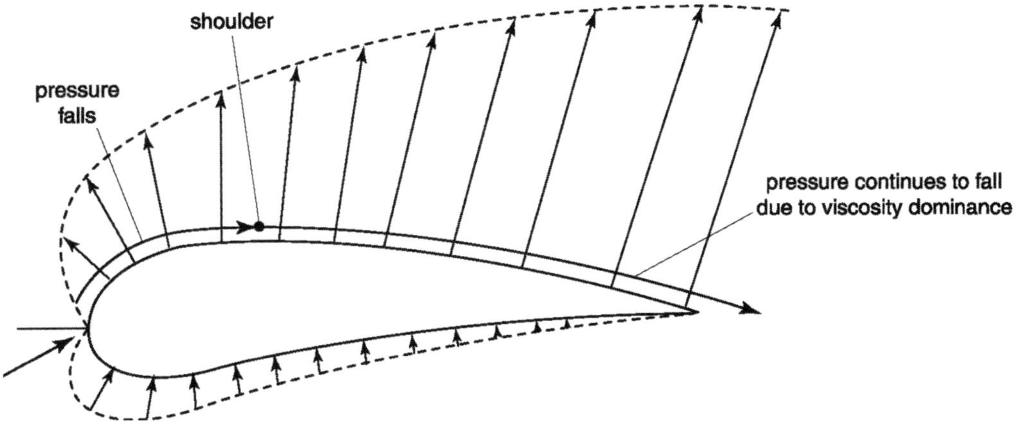

Fig. 1.8(b) Low Re flow

stress. This results in a greater drag than would be the case for the same airfoil at high Re. When the Re is high, there is greater mixing within the boundary layer due to inertia, and the flow "carries" itself along on the airfoil without separating. A given airfoil can have a higher drag at laminar flow than at turbulent flow, depending on the respective contributions of viscosity and inertia.

1.4 Airfoil Geometry

Airfoils are classified by the National Advisory Committee for Aeronautics (NACA) system. The same body later became known as the National Aeronautics and Space Administration (NASA). The terms used to describe the geometry of airfoil sections are shown in Fig. 1.4(a):

- *Leading edge*—This is the frontmost part of the airfoil. It is prescribed in terms of the leading edge radius, which is the radius of a circle that is tangent to the upper and lower faces of the airfoil, and with its center located on a tangent to the mean camber line drawn through the leading edge of the line.
- *Trailing edge*—The rearmost section of the airfoil is referred to as the trailing edge.
- *Mean-camber line*—This is the line that bisects the airfoil at any point of its travel between the nose and the tail of the airfoil.
- *Camber*—Also referred to as maximum camber, it is the maximum rise of the mean camber line from the chord line.
- *Chord line*—This is a straight line from the leading edge to the trailing edge of the airfoil section; it joins the ends of the mean camber line.
- *Airfoil thickness*—The thickness is defined as the maximum distance between the upper and the lower surface of the airfoil section.

Although developed for the aeronautical industry, airfoil theory has become valuable in the efforts to improve the performance of ground vehicles.

1.5 Flow Classification and Continuity: Steady and Unsteady Flow, Streamline

It is important to define some terms of the trade, since these come up from time to time in describing interactions taking place when an object goes through a fluid medium. The terms *laminar and turbulent* have been used to describe the flow over an airfoil. A more fundamental classification of flows identifies a flow as steady or unsteady. A turbulent flow is inevitably unsteady, and a laminar flow is typically steady. Fundamental classification of flows is as follows:

- Steady, uniform flow
- Steady, nonuniform flow
- Nonsteady, uniform flow
- Nonsteady, nonuniform flow

To define each of the types of flow listed, it is important to first establish the concept of streamline, path line, and streak line.

A streamline represents the path that a discrete fluid particle follows in the course of fluid flow. It is also referred to as flowline or line of flow. Streamlines are a family of curves that are instantaneously tangent to the velocity vector of the flow. They show the direction that a fluid element will travel in at any point in time. Streamlines are therefore solutions to the simultaneous equations

$$d\vec{x}/ds = \vec{v} \text{ or } dx_i/ds = v_i(x,y,z,t) \tag{1.1}$$

where

s is a parameter along the streamline, and

t is the temporal parameter, fixed while the equations are integrated.

If the components of the velocity are written as $\vec{v} = (u,v,w)$, and those of the streamlines as $\vec{x} = (x,y,z)$, then

$$\frac{dx}{u} = \frac{dy}{v} = \frac{dz}{w} \tag{1.2}$$

The resulting curves are the streamlines at the instant, t. Streamlines are calculated instantaneously. A streamline is defined by an Eulerian approach, as it gives the big picture of the flow at a particular time in the course of its travel.

A path line or particle path is the trajectory a fluid particle follows as it travels through space over time. A path line is defined by a Lagrangian approach, as it traces the movement of a particle with respect to time.

Path lines are defined by the simultaneous equations

$$d\vec{x}/dt = \vec{v} \text{ or } dx_i/dt = v_i(x,y,z,t) \tag{1.3}$$

where

$x_i = \alpha$, a given, at $t = 0$.

At point $\vec{x}$ the curve is parallel to the flow velocity vector $\vec{v}$, the latter being evaluated at the position $\vec{x}$ of the particle at time t.

A streak line is the curve traced out by a plume of smoke or dye that is continuously injected at a fixed point but does not diffuse. It presents the instantaneous picture of the positions of all particles that have passed through the point of injection.

$$d\vec{x}/dt = \vec{v} \text{ or } dx_i/dt = v_i(x,y,z,t) \tag{1.4}$$

where

$x_i = \alpha$, a given, at $t = t_1$, t_1 being such that $0 \le t_1 \le t_2$, with t_2 as the time of interest.

A streamline describes the line of motion of several particles at a given time, providing a snapshot of the flow, whereas a path line and a streak line depend on the full history of the flow; the path line traces a single particle in time and space and a streak line follows the particles that have passed through a point in space. In general, except for steady flow, the patterns of streamlines, path lines and streak lines differ for a given flow. It is difficult to observe streamlines in an experiment, but in a steady flow, because of the coincidence of all three flow forms, streak lines adequately serve to describe streamlines. Flow visualization of streamlines in steady state can be obtained in a smoke tunnel.

We will now revisit the flow classifications:

- *Steady, uniform flow*—Conditions do not change with position in the flow stream or with time. Because the flow is steady, streamlines do not change with time; subsequent particles follow the same path as the particles just ahead of them. An example of steady, uniform flow is the flow of air through the test section of a wind tunnel. The diameter of the test section is fixed, and speed is kept constant (at least for a given time interval).
- *Steady, nonuniform flow*—Conditions change from point to point in the stream but do not change with time. An example is flow in a tapering pipe. The velocity is constant velocity at the inlet and changes along the length of the pipe, be it divergent or convergent, toward the exit.
- *Nonsteady, uniform flow*—In this flow, the conditions at every point are the same at a given instant in time, but change with time. Because it is generally impossible to draw streamlines for unsteady flow unless time is considered, the term streamline flow is often used to describe steady flow, thereby differentiating steady from unsteady flow. An example of nonsteady, uniform flow is a pipe of constant diameter connected to a pump pumping at a constant rate that is then switched off.
- *Nonsteady, nonuniform flow*—A chaotic flow in which every condition of the flow may change from point to point and with time at every point. An example is air movement under the hood of a car in a windy condition.

1.5.1 Stream Tubes

When dealing with steady flow, particles cannot cross a streamline, and hence a set of parallel streamlines is called a stream tube. A stream tube is a bundle of streamlines that form a passage through which the fluid flows (Fig. 1.9). Unsteady flow typically results in change in energy level of the fluid between various points along the stream tube. Though the cross section through the streamlines at section 2 may be different in area and in shape from that at section 1, the same amount of fluid will pass per unit time through both sections because flow cannot occur across a streamline. There can therefore be neither gain nor loss of fluid between both locations or, for that matter, between any two locations in the flow as long as the flow is steady. The set of streamlines that bundle to provide the cross section thus shown form an artificial tube that is as effective as a tube with solid walls; hence, the term stream tube.

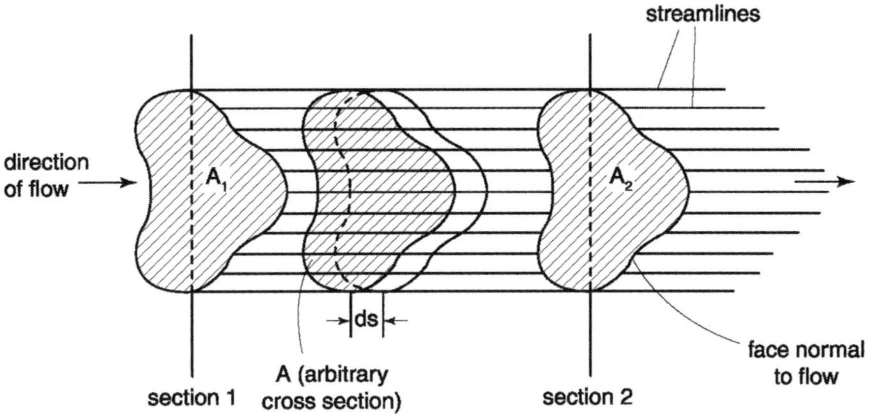

Fig. 1.9 Stream tube

1.5.2 Continuity

The mass of fluid of density ρ passing through an arbitrary cross section A over a distance ds in a time interval dt is

$$\rho A \frac{ds}{dt} = \rho A \mathrm{v}$$

If the mass flow per unit time is denoted by □, then

$$\dot{\mathrm{m}} = \rho_1 \mathrm{A}_1 \mathrm{v}_1 = \rho_2 \mathrm{A}_2 \mathrm{v}_2 \tag{1.5}$$

For an incompressible flow, the density is unchanged, and Eq. 1.1 reduces to

$$\mathrm{Q} = \mathrm{A}_1 \mathrm{v}_1 = \mathrm{A}_2 \mathrm{v}_2 \tag{1.6}$$

where Q denotes the volume flow rate.

Equations 1.5 and 1.6 are the equations of continuity. Continuity equations are fundamental to the study of fluid flow.

We can identify a streamline by the amount of flow that passes between it and some fixed point because by the definition of a streamline, there can be no flow added or taken away. The stream function ψ describes the amount of flow at a streamline. For a 2-D incompressible flow,

$$d\psi = \mathrm{v}dn$$

where v is the velocity along the streamlines, and *dn* is the distance between two stream lines.

When *dn* is broken to its components, we have it as

$$d\psi = \frac{\partial \psi}{\partial x}dx + \frac{\partial \psi}{\partial y}dy$$

or

$$d\psi = -\mathrm{v}_y dx + \mathrm{v}_x dy \tag{1.7}$$

where

$$\mathrm{v}_x = \frac{\partial \psi}{\partial y}$$

and

$$\mathrm{v}_y = -\frac{\partial \psi}{\partial x} \tag{1.8}$$

Equations 1.7 and 1.8 are referred to as stream function equations. When $d\psi = 0$, Eq. 1.7 reduces to the expression in Eq. 1.8, which is the equation of a streamline, confirming that lines of constant ψ represent streamlines. The numerical difference, $\psi_1 - \psi_2$, between the streamlines represents the flow rate per unit width passing between the streamlines [1-3, 1-4, 1-5].

1.5.3 Bernoulli's Equations

In Fig. 1.10, consider the forces on the elemental stream tube along the axis of the stream tube.

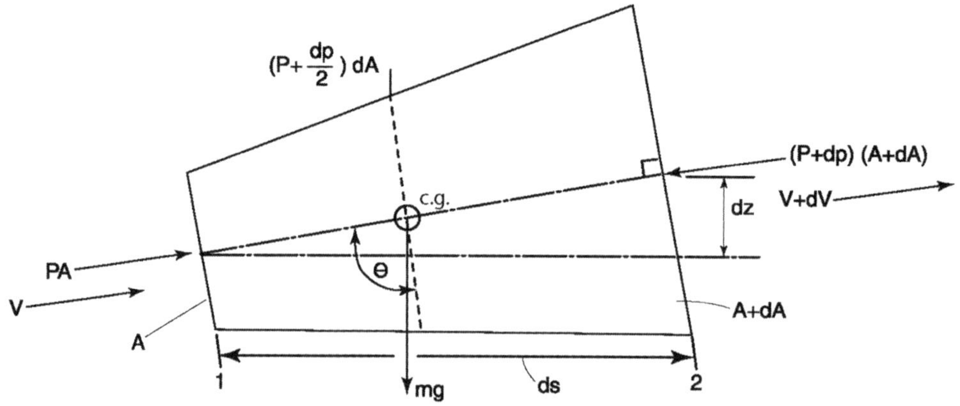

Fig. 1.10 Forces on elemental stream tube

$$ma = pA + \left(p + \frac{dp}{2}\right)dA - mg\cos\theta - (A + dA)(p + dp)$$

where

a is the fluid acceleration;

p is pressure at inlet;

A is surface area normal to fluid flow;

m is the mass of the elemental fluid; and

θ is the inclination of the line of action of the weight of the fluid to the axis of flow.

Ignoring the $dpdA$ terms, we obtain

$$ma = -mg\cos\theta - Adp \tag{1.9}$$

Substituting $a = \frac{1}{2}\frac{d(\text{v}^2)}{ds}$, and $\cos\theta = \frac{dz}{ds}$, in (1.9), and rearranging, we obtain

$$Adpds + mgdz + \frac{m}{2}d(\text{v}^2) = 0 \tag{1.10}$$

In systems where there may be friction or gas expansion, there may be thermal changes without the introduction of any external heat. This contribution can be accounted for by modifying (1.10) to

$$A(dp)(ds) + mgdz + \frac{m}{2}d(\text{v}^2) + mgdq = 0 \tag{1.11a}$$

The heat term dq is generally written in terms of the internal energy dU and the pressure work pdV resulting from the thermal change. So, substituting dq = dU + pdv in (1.11a) we obtain

$$A(dp)(ds) + mgdz + \frac{m}{2}d(\text{v}^2) + mg(dU + pdv) = 0 \tag{1.11b}$$

Equations 1.11(a) and (b) are the same and are valid for all adiabatic flows. From Fig. 1.10, $\frac{A(ds)}{mg}$ may be replaced in Eq. 1.11a by the specific volume, v, and, after dividing Eq. 1.11b throughout by the weight mg, we obtain

$$vdp + pdv + dz + \frac{1}{2g}d(\text{v}^2) + dU = 0 \tag{1.12}$$

Integrating Eq. 1.12 yields

$$pv + z + \frac{v^2}{2g} + U = C \tag{1.13a}$$

where C is a constant.

Writing U as $c_v T$ (from $dU = c_v dT$, where c_v is specific heat at constant volume, and T is absolute temperature), Eq. 1.13a becomes

$$pv + c_v T + z + \frac{v^2}{2g} = C \tag{1.13b}$$

Using the thermodynamic gas relationship $pv = RT$, where R is the gas constant, Eq. 1.13b is further refined to

$$pv(1 + \frac{c_v}{R}) + z + \frac{v^2}{2g} = C \tag{1.13c}$$

Employing the gas relationships $c_p - c_v = R$, and $\frac{c_p}{c_v} = \gamma$, Eq. 1.13c becomes

$$\frac{\gamma}{\gamma - 1}\frac{P}{\rho} + gz + \frac{v^2}{2} = C$$

The contribution of the gz term is generally negligible in flow around vehicles, so we can write the last expression without losing accuracy of result as

$$\frac{\gamma}{\gamma - 1}\frac{P}{\rho} + \frac{v^2}{2} = C \tag{1.14a}$$

or

$$c_p T + \frac{v^2}{2g} = C_1 \tag{1.14b}$$

Equations 1.14(a) and (b) are Bernoulli's equation for compressible flow. For an incompressible flow, which is the state of flow around ground vehicles, the internal energy as well as the work expansion terms are nonexistent. Therefore, Eq. 1.13a reduces to

$$p + \frac{\rho v^2}{2} = C_3 \tag{1.15a}$$

or

$$p_1 + \frac{1}{2}\rho v_1^2 = p_2 + \frac{1}{2}\rho v_2^2 \tag{1.15b}$$

The venturi and the pitot tube are examples of devices with design based on incompressible Bernoulli equation.

1.6 Boundary Layer

The boundary layer is the layer next to a body in a region within which the effects of viscosity are dominant. In Section 1.3 we had a preview to the concept of boundary layer. Here, we will give it an extended treatment.

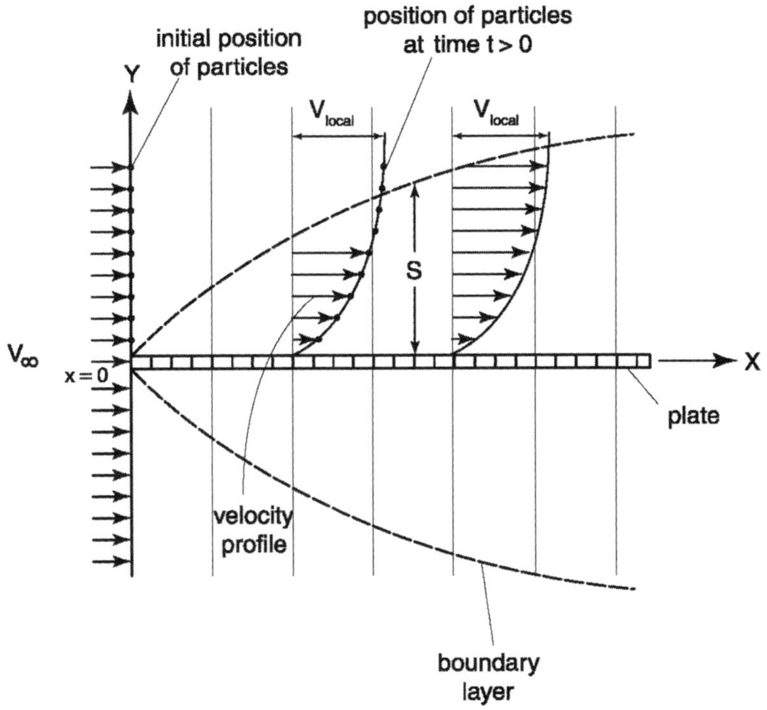

Fig. 1.11 Flow profile on a plate.

Consider the flow past a flat plate shown in Fig 1.11. At the fluid-plate interface, there is no relative motion between the fluid (air in this case) and the plate. This is referred to as the no-slip condition, and it prevails in all fluid flows except beyond the troposphere (>30 km), where the mean free path of air molecules is relatively large and the density is consequently low. Particles were released with uniform velocity at (x_o,t_o). The velocity distribution on the upper face of the plate at some point x along the plate is shown as the locus of the same particles at time t, later in the flow. The velocity profile begins from zero at the air particle in contact with the surface to the local free stream value as one moves farther away from the plate. This region of viscous influence is the boundary layer, and it is usually quantified by its thickness, the boundary layer thickness, δ. The value of δ depends both on the magnitude of the velocity and on the smoothness of the surface. At a given location from the incipience of the flow, as the velocity increases, the time slot for the viscous forces to fully manifest in response to the flow is less. Therefore the boundary layer thickness, which is a manifestation of the influence of the shear forces, reduces. The benefit of a smooth surface follows a similar reasoning. There is a direct correlation between the thickness of the boundary layer and the ability of the viscous forces in the boundary layer to dissipate the energy in the air stream. For a given point in the flow, a rough surface slows down the flow long enough for the viscous forces to dissipate the energy. A smooth surface, on the other hand, allows for a relatively faster movement of air molecules on the surface, thereby disallowing further build-up of the boundary layer. Hence, the boundary layer thickness is less.

When the flow lines are well organized, we have laminar flow. When there is a movement of fluid particles across the streamlines, even if the average velocity remains the same, the flow is no longer uniform, and it is referred to as turbulent flow (Fig. 1.12(a), (b)).

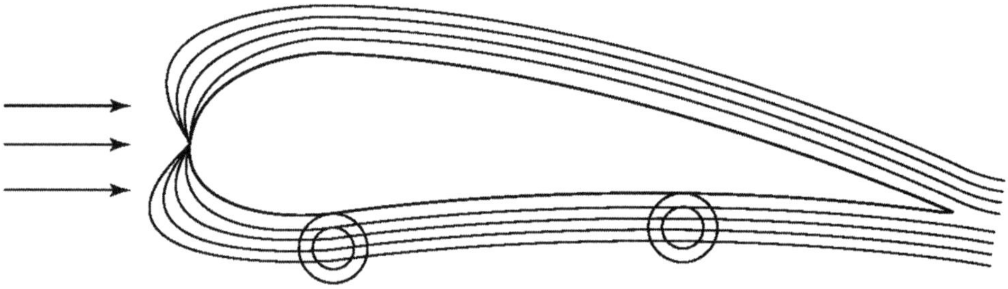

Fig. 1.12(a) Laminar flow

Fig. 1.12(b) Turbulent flow

The nature of the boundary layer determines a lot of the aerodynamic properties of a surface, including the skin-friction drag, the pressure drag, and the lift coefficient. There is exhaustive literature on methods to determine the boundary layer thickness [1-3, 1-4, 1-5, 1-6]. We will here provide a simplified approach using the method of dimensional analysis.

The boundary layer thickness, δ, is a function of the fluid density (ρ), viscosity (μ), the velocity of flow (v), and the distance along the plate (x). So we can write

$$\begin{aligned} \delta &= f(\rho, \mu, \mathrm{v}, x) \\ &= \mathrm{C}_1(\rho^a, \mu^b, \mathrm{v}^c, x^d) \end{aligned} \tag{1.16a}$$

where C_1 is a constant.

Applying mass length time (MLT) dimensions, and noting that the units of density, viscosity, velocity and distance are, respectively, kg/m^3, Ns/m^2 (= kg/(ms)), m/s, and m, the preceding expression becomes

$$L = C_1\left[\left(\frac{M}{L^3}\right)^a \left(\frac{M}{LT}\right)^b \left(\frac{L}{T}\right)^c (L)^d\right]$$

Equating exponents,

for L: $1 = -3a - b + c + d$

for M: $0 = a + b$

and, for T: $0 = -b - c$.

The simultaneous equations yield $a = c = -b$, and $d = 1 - b$.

When the values are substituted in (1.16a) we get

$$\delta_{laminar} = \mathrm{C}_1 x\left(\frac{\mu}{\rho \mathrm{v} x}\right)^b \tag{1.16b}$$

Prandtl's experimental work and Blasius's theoretical analysis, respectively, show that for laminar flow, C_1 is about 5.2 and b is 0.5. We therefore write Eq. 1.16b as

$$\delta_{laminar} = 5.2x\left(\frac{\mu}{\rho \mathrm{v} x}\right)^{0.5} \tag{1.17a}$$

The term in parenthesis is the inverse of Reynolds number (Section 1.1), so we can write Eq. 1.17a in a more compact and practical form as

$$\delta_{laminar} = \frac{5.2x}{\sqrt{\text{Re}}} \quad (1.17b)$$

In a similar manner, the skin friction coefficient, C_f, a nondimensional term critical in the determination of the viscous drag on a surface, is obtained. The local skin friction at position x along the flow on a surface is defined by

$$C_{fx} = \frac{D}{\frac{1}{2}\rho v^2 A} \quad (1.18)$$

where

D is the drag on the surface;

A is the area of the surface experiencing the drag;

$\frac{1}{2}\rho v^2$, the dynamic pressure is the nondimensionalizing term.

The average skin friction coefficient on a surface is

$$C_{f_{laminar}} = \frac{1.328}{\sqrt{\text{Re}}} \quad (1.19)$$

A turbulent boundary layer has a greater capacity to remove energy from the free stream than does a laminar layer due to the momentum interchange with the flow outside the boundary layer. This capacity results in higher energy in the turbulent region. It also results in a higher skin friction drag, as more energy is taken from the free stream to perpetuate the boundary layer [1-7].

For turbulent boundary layer on a flat plate

$$\delta_{turbulent} = \frac{0.37x}{\text{Re}^{0.2}} \quad (1.20)$$

and

$$C_{f_{turbulent}} = \frac{0.074}{\text{Re}^{0.2}} \quad (1.21)$$

The transitional stage between laminar and turbulent flow has been found to fit the following relationship

$$C_{f_{transition}} = \frac{0.074}{\text{Re}^{0.2}} - \frac{1700}{\text{Re}} \quad (1.22)$$

The transition region is so small that in general, the boundary layer thickness for this region is split into the thickness for laminar and turbulent flows. The trend of skin friction variation with Re is shown in Fig. 1.13.

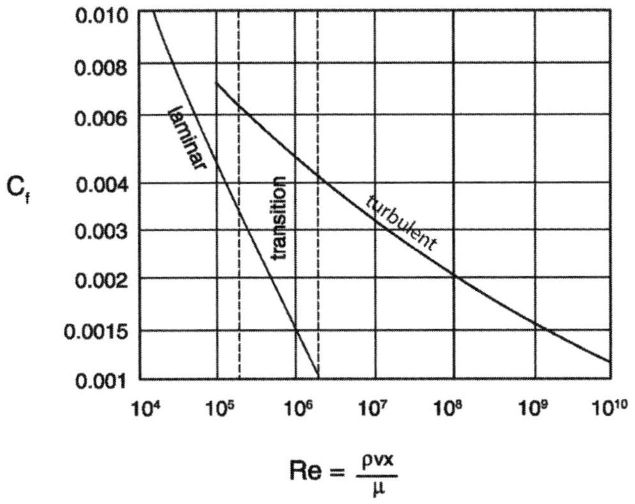

Fig. 1.13 Trend of skin friction coefficient on a smooth, flat plate

This parameter, C_f, is very important in the determination of the drag on a surface, such as the body of a car. As indirectly shown in (1.18), the friction drag, D_f, on a surface is defined as

$$D_f = \tfrac{1}{2}\rho v^2 C_f A$$

The turbulent boundary layer is not all turbulent. There is still an underlying, albeit thin, layer of laminar flow on the surface of the plate due to the ever-present viscosity on every surface, no matter how well polished (Fig. 1.14).

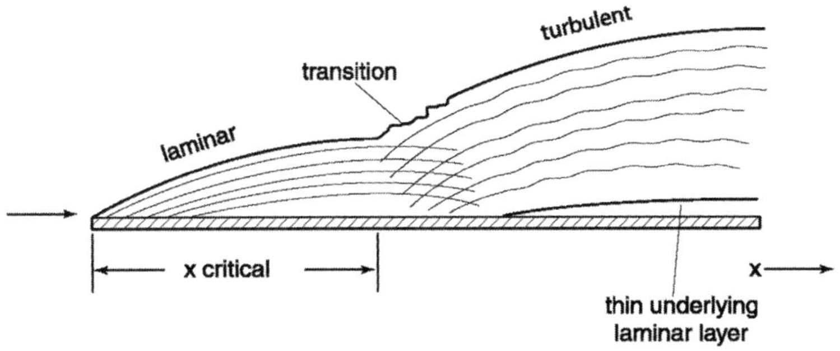

Fig. 1.14 Boundary layer on a plate

Because the skin friction is higher in turbulent flow than in laminar flow, there is the temptation to delay the occurrence of turbulence so that a streamlined, minimal-friction flow can be enjoyed over a greater length of a surface. This is usually attempted by keeping the Reynolds number low. But, as discussed in Section 1.3, this effort may backfire, resulting in a higher drag either because the profile, such as the side of a car, has become too long to maintain the benefit of continuous laminar flow or the curvature that the flow encounters is too sharp. On the other hand, one may want to quicken the transition to turbulent flow; this is common in drag race cars with vortex generating devices such as roughening the surface at a location in the profile. The benefit is maintaining the flow attachment to the surface, thus avoiding separation or at least reducing the probability of separation. The penalty in skin friction drag needs to be weighed against the benefits before embarking on this path.

1.7 Airfoil Theory

A brief introduction to the airfoil was made in Section 1.4. An airfoil is a sectional view of the symmetric plane through a teardrop. It is also a 2-D cross section of a wing. This last description is traditionally employed in the analysis of an airfoil because of the practical applications. The primary step in the development of an airfoil is the establishment of the mean camber line and the distribution of the thickness. If we apply this philosophy to car design, we would first establish the longitudinal profile of the car and the height distribution of the car from the grill to the rear bumper location.

The effect of an airfoil is to cause the flow to be faster on the upper surface of the airfoil and slower on the lower surface. The consequence of such a flow profile is a resultant force acting to push the airfoil in the directional sense of the upper surface; this we call lift. Airfoil theory and its integration, wing theory, were developed for aircraft aerodynamics. Consistent with that background, this topic and wing theory are treated with a perspective on aircraft wing. There is no conflict in the use of this information for ground vehicles because improved lift, the primary benefit of the understanding of airfoil theory, becomes improved downforce when mirrored about the horizontal plane.

There are various schools of thought on airfoil lift theory. They range from the attack angle (or Newtonian) theory to circulation theory. In between, we have airfoil shape (or path length) theory, streamline curvature theory, and vortex theory. There is no unified airfoil theory, but there is enough understanding of the flow around an airfoil to design effective wings and blades. Without defining a school of thought, a plausible explanation will be provided here.

Consider Fig. 1.15(a) and (b). Experiments have shown that the flow lines are closer together in the vicinity of an obstacle when the fluid passes the obstacle, in this case a smooth sphere, as shown in Fig. 1.15(a). This is so even when the fluid is unbounded, as in a wide stream or a river. This is consistent with boundary layer theory. Experiments have also shown, as in Fig. 1.15(b), that the flow lines around the smoother part of the obstacle are closer together than around the relatively blunt side. (This is easily observed in flow over rocks in a riverbed). Figure 1.16(a) can be considered as an extension of Fig. 1.15(b). The flow over the upper face in Fig. 1.16(a) is, as expected, more packed. This implies a higher velocity on the upper surface of the airfoil, leaving a higher dynamic pressure to push up the airfoil at the fore-section. But this does not explain the lift in a symmetrical airfoil. To explain that, we note that airfoils are generally positioned at an angle α, the angle of attack to the flow stream (Fig. 1.16(b)).

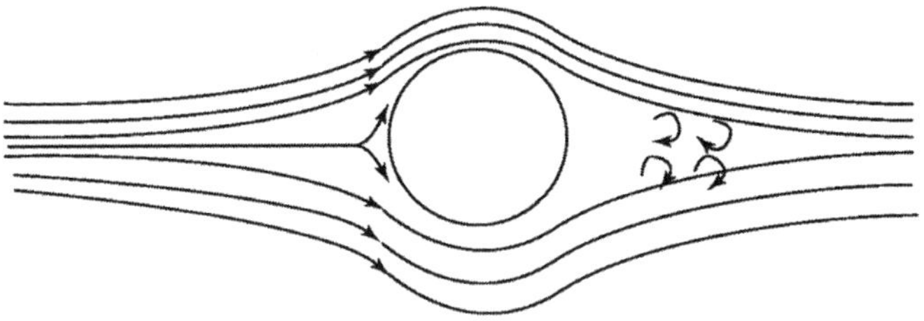

Fig. 1.15(a) Fluid flow over a sphere

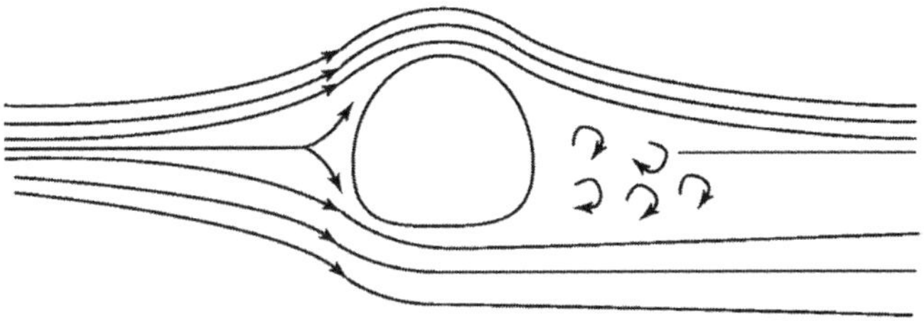

Fig. 1.15(b) Fluid flow over a partly spherical surface

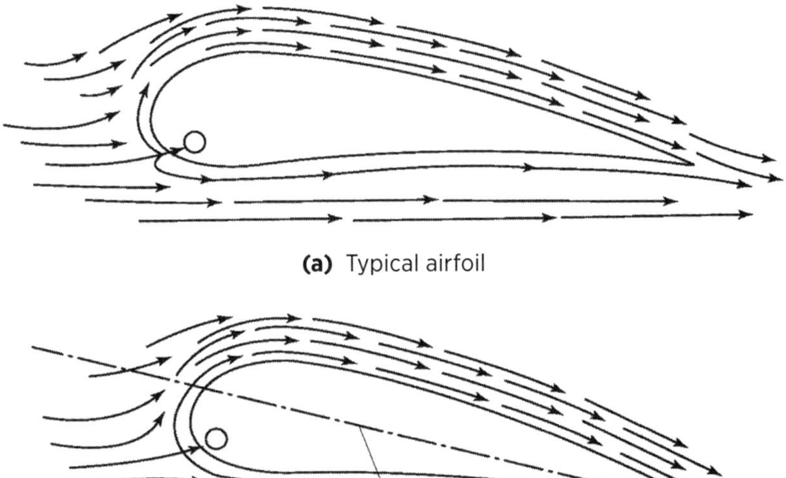

(a) Typical airfoil

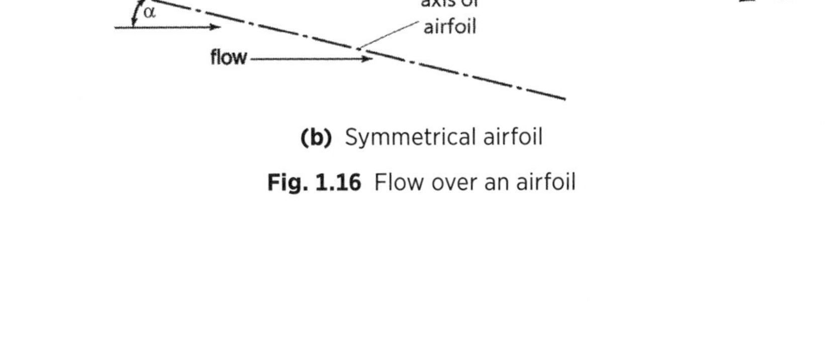

(b) Symmetrical airfoil

Fig. 1.16 Flow over an airfoil

A fluid particle traveling to the left of point O will turn leftward and follow the curvature of the airfoil. Because it turns to the left, against the approaching flow, its velocity is reduced, thereby causing an increase in the dynamic component of the pressure. Because the static component of the pressure is unchanged, the total pressure increases. Additionally, the interaction with the head-on fluid particles causes an increase, even if slight in molecular action at the location, and this adds to the pressure. As the fluid particle follows its constrained path around the nose of the airfoil and down the curve of the upper surface, it speeds up (recall Fig. 1.15(a), (b)). This results in a lower pressure on the leading edge of the airfoil up to a region between the camber (which defaults to the leading edge in a symmetric airfoil) and the airfoil maximum thickness location along the longitudinal axis. Airfoil lift is attributable to these combined effects. The greater the camber (up to about 12.5% of the chord length), the greater is the lift experienced by the airfoil. Airfoil thickness also contributes to lift, albeit slightly. A thicker airfoil improves lift for subsonic flow. Certainly, the greater the angle of attack (up to about 17°), the greater is the lift. Figure 1.8(a) shows the typical pressure distribution on an airfoil. In Fig. 1.17, we see how pressure distribution on an airfoil varies with the angle of attack.

There are perhaps four main parameters used in quantifying forces on an airfoil: the lift coefficient, drag coefficient, moment coefficient, and pressure coefficient.

$$\text{Airfoil lift coefficient, } c_l = \frac{L}{\frac{1}{2}\rho v_\infty^2 A}$$

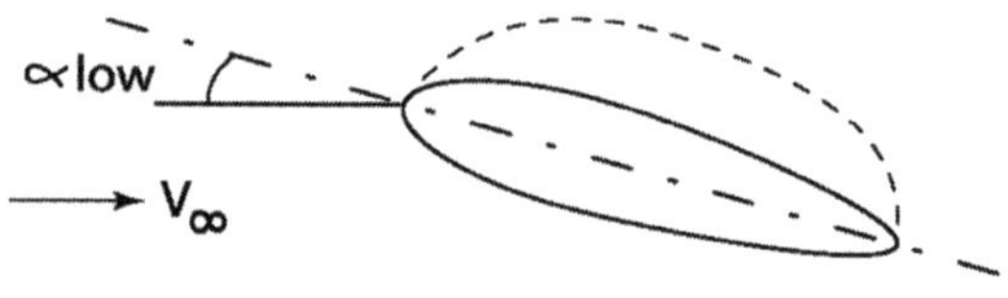

Fig. 17(a) low α

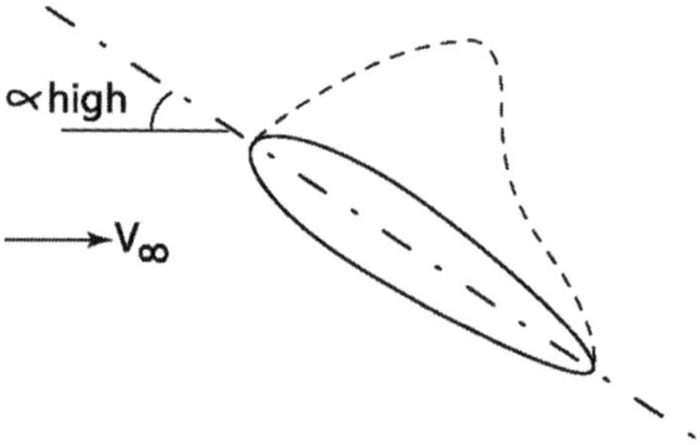

Fig. 17(b) high α

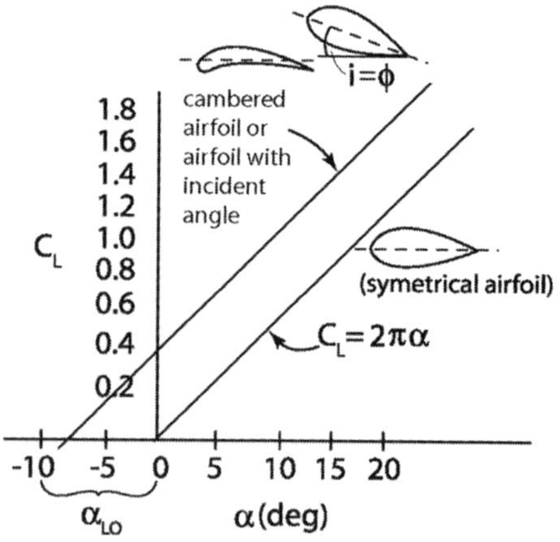

Fig. 17(c) C_L vs. α

Fig. 1.17 Pressure and C_L variation with α

where

L is the lift on a 2-D wing;
A is the projected area in the chord (or lateral) plane of the wing;
v_∞ is the local free stream velocity.

Similarly

$$\text{airfoil drag coefficient, } c_d = \frac{D}{\frac{1}{2}\rho v_\infty^2 A}$$

$$\text{and the pressure coefficient, } C_p = \frac{p - p_\infty}{\frac{1}{2}\rho v_\infty^2}$$

The lift and the aerodynamic drag are obtained by the integration of the pressure distribution over the wing surface area. From statics, if the distributed forces are converted to a concentrated force, there would be an added moment. The location of the representative aerodynamic force is referred to as the center of pressure (c.p.) (Fig. 1.18) Although the c.p. usually falls around the aerodynamic center (a.c.), it is not as a rule coincident with the a.c.; it varies with wing loading. The a.c. is the point about which the pitching moment, c_m, is independent of α, the angle of attack. It is as a rule taken as a quarter-chord length (c/4) from the leading edge. The c.p. is usually compared with the center of gravity (c.g.) in vehicle performance analysis, whereas the a.c. is significant in airfoil selection.

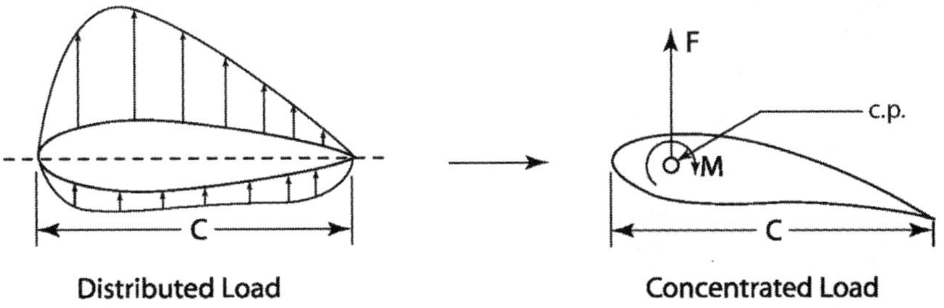

Fig. 1.18 Center of pressure (c.p.).

The pitching moment coefficient is defined as

$$c_m = \frac{M}{\frac{1}{2}\rho v_\infty^2 Ac}$$

The lift coefficient is often expressed in terms of the angle of attack, thus for a symmetric airfoil,

$$c_l = 2\pi\alpha$$

and, for a cambered airfoil or one fixed at an angle of incidence $i = \alpha_{L0}$

$$c_l = 2\pi(\alpha + \alpha_{L0})$$

The angle of incidence, *i*, which the airfoil makes with the reference line, is a reference measurement, fixed, as it defines the inclination of the airfoil (a wing, for example) to a level ground.

The drag coefficient is usually expressed as a fraction of the lift coefficient. This helps to rate an airfoil readily since the goal is to maximize the lift coefficient and correspondingly minimize the drag coefficient.

1.8 Lift, Drag, Side Force, and Moment

There are many forces acting on an object as it moves through a fluid medium. They include electrical, magnetic, gravity, pressure, and friction. In everyday applications for ground vehicles, only pressure and friction forces are of significance. The pressure forces function mostly to produce lift, and the friction forces cause drag. Figure 1.7a shows an airfoil with the resultant force broken into its two components, lift (L) and drag (D). The chord line is inclined at an angle α, the angle of attack, to the relative wind. The angle of attack defines the attitude of the airfoil's chord line with respect to the head-on air flow. The lift acts normal to the relative wind through the c.p. of the airfoil while the drag is parallel to the relative wind, acting to retard motion. The c.p. is (as pointed out in Section 1.7) the location of the resultant of all forces acting on the airfoil. This is significant, as we will see in the treatment of vehicle performance. The relationship between the c.p. and the c.g. determines to a large extent the dynamic stability condition of a vehicle.

Because drag resists motion, a certain amount of power is needed to overcome the retarding action of drag. Drag may therefore be considered as the penalty for producing motion and lift (or, for ground vehicles, down force). The lift-to-drag (L/D) ratio is an indication of the efficiency of the airfoil.

In general, when an object, like a car, moves through the air, the forces that act on the object are determined by the relationship between the free stream velocity, the air density, the viscosity, the surface smoothness of the object, the (frontal) area and the shape of the object, and the inclination of the object to the approaching air flow. The equations of lift, drag, and moment are determined in terms of the listed influences, experimentally or theoretically.

As in the case of the airfoil, the coefficients of the forces and the coefficients of moment are useful tools both for a case-by-case analysis and for comparison of different designs. Unlike in the case of the airfoil, these parameters represent the coefficients for the full 3-D flow, so, *A* is the area of the full wing planform.

Forces:

$$\text{Lift coefficient, } C_L = \frac{L}{\frac{1}{2}\rho v^2 A}$$

$$\text{Drag coefficient, } C_D = \frac{D}{\frac{1}{2}\rho v^2 A}$$

$$\text{Yaw coefficient, } C_Y = \frac{Y}{\frac{1}{2}\rho v^2 A}$$

where Y is the yaw or the side force.

Moments:

$$\text{Coefficient of pitching moment, } C_M = \frac{M}{\frac{1}{2}\rho v^2 Al}$$

$$\text{Coefficient of rolling moment, } C_R = \frac{R}{\frac{1}{2}\rho v^2 Al}$$

$$\text{Coefficient of yawing moment, } C_N = \frac{N}{\frac{1}{2}\rho v^2 Al}$$

where, as shown in Fig. 1.19,

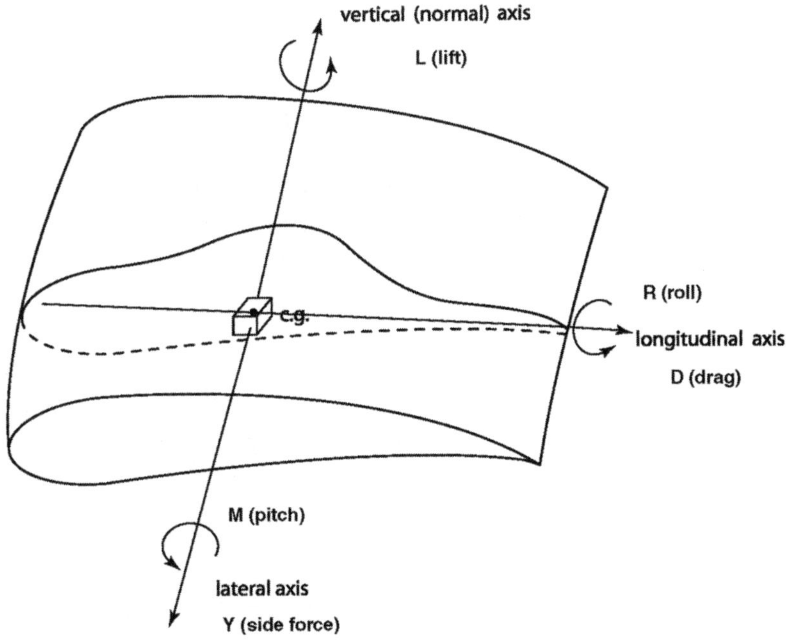

Fig. 1.19 Coordinate system for loads and moments

M is the pitching moment, about the lateral axis;

R is the rolling moment, about the longitudinal axis;

N is the yawing moment, about the vertical axis;

l is a characteristic length, usually the wheelbase.

Aerodynamic drag of a vehicle manifests in the following forms.

1.8.1 Form Drag

Every object has a form drag. Also referred to as pressure drag, form drag exists because a shape exists. It is created by the basic shape of the object. Air is forced around the object as the air moves opposite to the direction of motion of the object. The size and shape of an object determine the magnitude of the form drag. Exhaustive tests have shown that the least form drag is obtained when the body has an airfoil shape, and the highest is when the body is wholly blunt (i.e., a flat surface directed normal to the head-on air flow).

In real life, objects are neither fully airfoil nor wholly blunt. Aerodynamic designers aim for the optimum low-drag shape, which is one that is as close to an airfoil as design constraints will permit for a given mission specification. This is done by contouring the shape to minimize deflection around the body and to reduce the turbulence at separation and at reattachment.

The form drag of a car is typically the dominant drag, at about 80%, compared with the skin friction drag, which is about 12%.

1.8.2 Skin Friction Drag

Skin friction or surface friction is the shearing effect created in the boundary layer of a body. The longer the body, the greater is the build-up of the boundary layer (since it takes time to build up boundary layer), and the higher is the surface friction. Because viscosity is the direct cause of skin friction drag, it is also called viscous drag. Profile drag is the sum of the skin friction drag and the form drag.

1.8.3 Interference Drag

Protrusions, even small ones such as mirrors, door handles, head lights, tail lights, and hood ornaments, interfere with the general flow around a car. Apart from the direct drag of the protrusion, there is the interference drag that is a result of the interaction of the main body drag with the protrusion drag. Interference drag is usually determined in a wind tunnel test. It is obtained by testing the bulk of the body, the protrusion, and the combination separately. Wind tunnel tests have shown that the drag from the test of the combination is more than the sum of the drag from the separate tests of the bulk and the protrusion. This difference is referred to as the interference drag.

The significance of interference drag is in the importance of parts integration for an aerodynamically suitable whole. Although it is important, from an aerodynamic perspective, to design a component so as to minimize the drag incurred by that component, it is as important to design the system so as to have a minimal system drag incursion. The farther the component is from the main body, the less is the interference drag contribution.

Parasitic drag is a term used to describe all the aerodynamic components except the induced and the wave drag. In general, the sum of the form drag, the skin friction drag, and the interference drag is equated to the parasitic drag.

$$\text{Profile drag} = \text{Skin friction drag} + \text{Form drag}$$
$$\text{Parasitic drag} = \text{Skin friction drag} + \text{Form drag} + \text{Interference drag}$$
$$= \text{Profile drag} + \text{Interference drag}$$

1.8.4 Induced Drag

Induced drag results from vortices, which cause aerodynamic lift. Unlike parasitic drag, induced drag decreases with vehicle speed (Fig. 1.20). Every object experiences induced drag, but it is more pronounced in airplanes and, to a lesser extent, in ground vehicles.

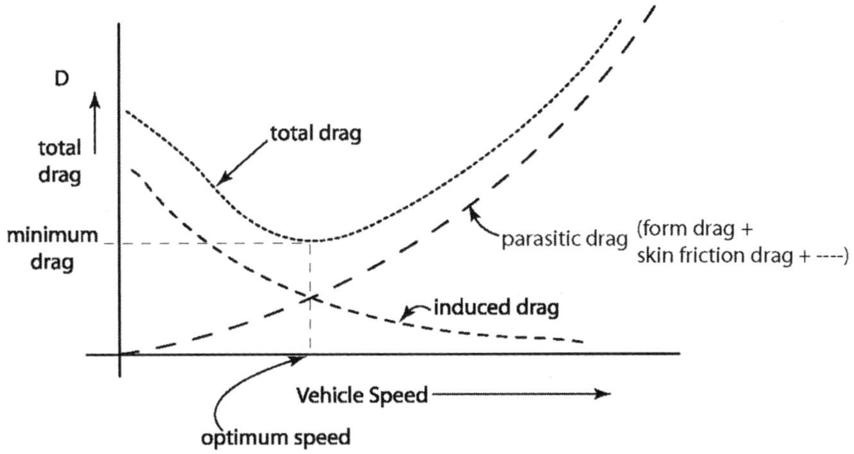

Fig. 1.20 Drag variation with vehicle speed

1.8.5 Wave Drag

Wave drag is the result of the shock waves formed around a vehicle, typically a high-speed flight vehicle. A sudden, powerful retarding force accompanies the shock wave that forms ahead of a vehicle as it approaches the speed of sound. It is unlikely that a ground vehicle experiences wave drag, because it takes a speed of about 608 mph (980 km/h) to have a shock wave.

1.9 Sources of Drag in Ground Vehicles

Apart from the obvious sources—the car profile, protrusions such as the side mirror and other attachments to the main body—of drag, there are contributions from internal flow (chapter 12), rotating wheels, and the underside. It is typically the combined effect of all of these components that is recorded when the aerodynamic drag of a vehicle is stated.

Summary

The benefits of reduced aerodynamic drag include fuel efficiency, improved vehicle performance, and increased passenger comfort.

The boundary layer region is a region of retarded flow. Outside the boundary layer, the flow is free of viscosity and is referred to as streamline flow. A streamlined airfoil shape is ideal for the near-elimination of boundary layer separation.

Reynolds number is a common term used to express the relationship between the determining parameters of a boundary layer flow. Reynolds number is the ratio of inertia term to the viscous term of a flow. At low speeds, the viscous force outweighs the inertia force, resulting in a low Reynolds number. At high speeds, the inertia force is superior to the viscous force; hence, a high Reynolds number.

At low Reynolds number, the drag is proportional to the speed and to the effective size of the object and inversely proportional to the viscosity. The flow density does not have a significant effect in determining the Reynolds number at low Reynolds number flows. At high Reynolds number, however, the density is a strong determinant of the Reynolds number. The Reynolds number varies with the square of the speed and of the object effective size, along with the flow density.

The main types of drag experienced by ground vehicles are form drag and skin friction drag. Form drag accounts for about 80% of ground vehicle drag, and skin friction drag claims about 12%. The rest is attributed to interference drag and induced drag.

$$\text{Profile drag} = \text{Skin friction drag} + \text{Form drag}$$
$$\text{Parasitic drag} = \text{Skin friction drag} + \text{Form drag} + \text{Interference drag}$$
$$= \text{Profile drag} + \text{Interference drag}$$

Induced drag is caused by vortices, which cause aerodynamic lift. It is an important parameter in vehicle stability and performance evaluation. Unlike parasitic drag, induced drag decreases with speed.

Key Terms

Airfoil: A surface designed to assist in lifting or aerodynamically controlling an object by making use of the air currents through which the object moves.

Boundary layer: A region of retarded flow around an object; it is influenced by the shape of the object.

Center of pressure: The location of the resultant of all forces acting on the airfoil.

Density: The amount of matter (i.e., mass) occupying a given volume of a substance.

Drag: The aerodynamic force that inhibits motion through the air.

Force: The influence that moves an object from rest or that changes its rate of motion.

Laminar flow: A flow in which the fluid particles move downstream in smooth and regular trajectories with almost no mixing between different layers of the fluid.

Lift: A component of the body force that is perpendicular to the flow direction of the oncoming air.

Moment: The tendency of a force to twist or to rotate an object.

Reynolds number: Ratio of inertia forces to viscous forces on an object.

Steady flow: A flow in which all the conditions at a given point in the flow are constant with respect to time.

Stokes flow: Low Reynolds number flow.

Streamline: In a flow, the streamline is a path traced out by a massless particle moving with the flow.

Turbulent flow: A flow in which there is mixing and interchange of mass and momentum between different streamlines; turbulent flows typically result in higher drag than laminar flows.

Unsteady flow: A flow in which the conditions at a given point in the flow vary with changing time.

Viscosity: The property of a fluid that resists the force tending to cause the fluid to flow.

Review Questions

1. List three benefits of reduced aerodynamic drag on ground vehicles.

   Apart from aerodynamic drag, what other forms of drag does a ground vehicle experience?

2. Define boundary layer. What is the difference between laminar boundary layer and turbulent boundary layer? How does boundary layer generally vary with speed?

3. Airplanes are designed for high aerodynamic lift coefficients. But zero or a negative lift coefficient is the design goal for ground vehicles. Explain the difference in these design goals.

4. What type of vehicle design is likely to be less stable at high speeds, all other factors being constant, a high-nosed car or a car with a lowered front end? Explain.

5. What is interference drag? List three sources of interference drag in ground vehicles.

Important Questions/Application Questions

1. From a strictly aerodynamic standpoint, what kind of landscape might be the best to attempt speed records for ground vehicles?

2. Why is laminar flow not always desirable for a ground vehicle?

3. Compare the body features of a high-volume production car to that of a performance car. Which of the features do you think are aerodynamically motivated?

Exercises

1. What is the Reynolds number at the trailing edge of a horizontally plate with a 15 in. chord? The plate is at the leading edge of a race car moving at 180 mph at standard sea level condition.

2. What is the total skin friction drag on a low-roof car that is 4m long and 1.8 meters wide if the car is traveling at 140 km/h? The transition Reynolds number for the car surface is 750,000. Assume the car to be a flat plate from the leading edge to the trailing edge. Do not include underbody flow.

3. Do Exercise 2 with the flow assumed to be (a) completely laminar, (b) completely turbulent. Compare the results in (a) and in (b), respectively,

with the result in Exercise 2 and comment on the assumptions made in (a) and in (b).

Projects

1. Design an experiment to determine the projected frontal area of your car. Determine also the projected plan and the projected vertical area. Do the same for five other cars of different makes.

   Find the design weight (not gross weight) of each car. What kind of relationships are you able to draw from the data?

2. Take an 8ft by 4 ft sheet of metal or smooth plywood. Put pressure tabs at about 2 in. intervals along the longitudinal center.

   Case 1: Keep the sheet metal flat. At the leading edge, direct forced air along the length of the sheet metal.

   - Measure the pressure along the tabs.
   - Repeat for nine more speed settings.

   Case 2: Allow the sheet metal to drop naturally under its own weight, keeping a support structure underneath in the midsection region. Repeat the steps in Case 1 (**Use the same speed settings as in Case 1**).

   - For each case, discuss the pressure variation with the leading edge speed.
   - For each speed setting, compare the pressures of both designs.
   - Discuss your results.

References

1. Shapiro, A. H., *Shape and Flow–the Fluid Dynamics of Drag*, Anchor Books, Doubleday & Company, Inc.: New York, 1961.
2. Kuethe, A. M., and Schetzer, J. D., *Foundations of Aerodynamics, 2nd ed.*, John Wiley & Sons, Inc., 1959.
3. Dommasch, D. O., Sherby, S. S., and Connolly, T. F., *Airplane Aerodynamics, 4th ed.*, Pitman Publishing Corporation, 1967.
4. Streeter, V. L., and Wylie, E. B., *Fluid Mechanics, First SI Metric Edition*, McGraw-Hill Ryerson Limited, 1981.

5. Massey, B. S., *Mechanics of Fluids, 5th ed.*, Van Nostrand Reinhold (UK) Co. Ltd., 1983.

6. Sabersky, R. H., Acosta, A. J., and Hauptmann, E. G., *Fluid Flow: A First Course in Fluid Mechanics*, 2nd ed., Macmillan Publishing Co., Inc., 1971.

7. Schlichting, H., *Boundary Layer Theory*, McGraw-Hill, 1960.

Chapter 2
Noise and Vehicle Soiling

Noise is undifferentiated sound. To contrast musical sound to noise in acoustic terms, we may describe musical sound as resulting from regular, periodic vibrations, whereas noise results from non-periodic vibrations. The sources of noise are fundamentally different from the sources of drag, so it is possible and indeed often is the case that there is no correlation between the wind noise experienced in a car and the aerodynamic drag of that car. Typical sources of wind noise in a vehicle are the pillars (especially A-pillar), the windshield, gaps in doors, and cracks in seals, whereas vehicle aerodynamic drag is due to the overall flow over the vehicle. Although there is no established correlation between vehicle aerodynamics and vehicle noise, it is conceivable that, at flow separation and in turbulent regions of flow, noise is generated [2-1].

Aerodynamic noise is an unsteady flow phenomenon, with pressure fluctuation in a certain frequency range. It is mostly caused by the separation of flow from and reattachment of flow to the surface. It can be abrupt or consistent, depending on the ambient air flow condition and on the design of the obstruction. Special aerodynamic tunnels can measure aero-acoustics, the aerodynamic sound created by the turbulent currents and its interaction with the surface of the vehicle. There are a few publications that provide detailed theoretical treatment of aero-acoustics [2-2, 2-3, 2-4, 2-5, 2-6, 2-7]. The goal of the treatment of acoustics here as it relates to ground vehicles is to understand the sources and forms of noise generated by vehicle-environment interaction and to design for noise attenuation. A level of noise is always experienced by vehicle occupants, but it becomes significant at frequencies that are higher than 600 Hertz and at vehicle speeds above 50 mph (80 km/h). Noise generation would be insignificant without the reception of noise, because the irritation or attention that noise creates is based on the value or the form of the reception. The receiver is either the vehicle occupant or a medium such as a microphone (Fig. 2.1).

Fig. 2.1 Microphone positions to measure sound pressure level (SPL) for driver and front passenger (reproduced with the permission of SAE International)

Noise is quantitatively expressed in decibels (dB), a sound level unit. The decibel scale is a convenient unit to describe the ear's response to sounds. A more direct measure is the intensity (I). The energy associated with a wave is proportional to its amplitude squared. The amplitude of a sound wave is related to the gauge pressure of its compressions and rarefactions, so the energy carried by a sound wave varies with its overpressure (compression) and underpressure (rarefaction). Wave energy spreads out from a source—the wave source—and is defined at a given location by the intensity, I, given by

$$I = \frac{E/t}{A}$$

where

E/t is energy per unit time; and

A is the area of the wave spread.

Since energy per unit time is power, intensity is power per unit area.

$$I = \frac{power}{A}$$

From the last expression, for sound intensity, the discomfort—sometimes damage—that noise causes can be explained. Noise, typically a very intense sound, carries energy into the human ear faster than the ear can safely dissipate the energy. The eardrum is therefore forced to vibrate at a large amplitude that can damage the eardrum and the hearing mechanisms in the ear.

Intensity is commonly expressed in watts per meter squared (W/m^2). The human ear can safely detect sounds of an intensity range from 10^{-12} W/m^2 to 1 W/m2.

The ear's response to sound is logarithmic rather than linear. For this reason (at least, in part) the decibel scale, a logarithmic form of intensity, is used.

$$\beta = 10\log_{10}(I/I_o) \quad (2.1)$$

where

$I_o = 10^{-12}$ W/m^2, the threshold of hearing at a frequency of 1000 Hz.

Table 2.1 shows the sound levels and intensities of some common sounds.

Table 2.1 Sound Levels and Intensities for Various Activities

Sound Level (dB)	Intensity (W/m^2)	Activity
0	10^{-12}	Threshold of hearing at 1000 Hz
40	10^{-8}	Soft music
60	10^{-6}	Normal conversation
70	10^{-5}	Busy traffic
80	10^{-4}	Loud radio
90	10^{-3}	Inside subway train
120	1	Loud rock concert indoors, pneumatic chipper at about 3 m; threshold of pain, damage to ear in minutes
160	10^{4}	Eardrum bursts

Example 2.1

a. For a sound intensity of 10^{-8} W/m^2, what is the sound level in decibels?

b. Given a sound level of 50 dB, find the intensity in watts per meter squared.

Solution: Use Eq. 2.1

a. $\beta = 10\log_{10}(I/I_o)$

$$\beta = 10\log_{10}\left(\frac{10^{-8}\ \text{W/m}^2}{10^{-12}\ \text{W/m}^2}\right)$$

$= 10\log_{10}10^4$

$= 40$ dB. This is considered reasonable. It is a typical value for soft music and for the average home.

(continues)

b. $50 = 10 \log_{10}(I/I_o)$

$I/I_o = 10^5$

$I = 10^5$

$I_o = 10^5 (10^{-12} \text{ W/m}^2)$

$I = 10^{-7} \text{ W/m}^2$, which is the sound intensity in a typical office setting.

Noise causes passenger discomfort and can have a direct negative impact on vehicle sale. It is therefore important to understand, early in the design phase, the correlation between a given vehicle geometry and the resulting in-cabin noise.

2.1 Sources of Noise

The main sources of noise in ground vehicles include the vehicle body, windshield and side glasses, powertrain, HVAC system, environment, and tires. Figure 2.2 shows a nearly exhaustive list of inherent sources of noise in and around ground vehicles.

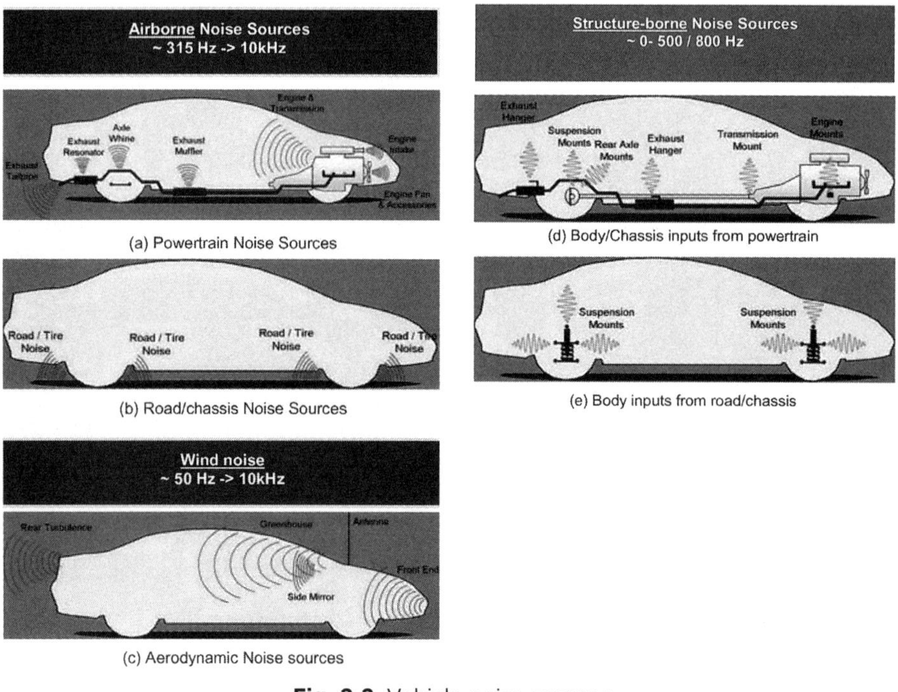

Fig. 2.2 Vehicle noise sources
(reproduced with the permission of SAE International)

At highway speeds, aerodynamic noise dominates the overall cabin noise level [2-8]. The relative importance of aerodynamic noise has increased substantially, in part because of the strides made in the reduction of the noise contribution from the engine, powertrain, air duct system, and tires, and in part because of the consumer's appetite for greater in-cabin comfort. The attention given to aerodynamic noise is expected to increase as we experience an increase in vehicles powered by the quieter electric and hybrid engines.

The primary sources of aerodynamic noise are the vehicle body and the side glasses.

2.1.1 Vehicle Body

Body noise is generated when there are movements of vehicle parts that result in metal-to-metal impact. It consists of bounce, shake, and rattle movements, in general. The reaction between tire and road surface creates dynamic forces that are transferred into linkages in the vehicle structure. The road condition is a main determinant in the severity of this type of noise. Well-paved roads have even surfaces and are much less likely to cause a shock on the tire, so there is minimal dynamic load for the wheel system to transfer to the linkages; whereas vehicles vehicles traveling on uneven road surfaces tend to gallop a lot, thereby transferring tremendous dynamic forces to the linkages in the vehicle structure.

The air pressure on the auto body is another contributor to vehicle noise. Although both the shear pressure and the direct pressure due to compressive wind load on the auto body cause noise, the latter form has a more significant contribution and is usually in the form of side force hitting both the side glass and the body. The absorbent quality of the skin material used in making the car body and the thickness of the material are important determinants of the ability of the car body to "soak" and to deflect the noise that reaches it. For a given material or composition, the thinner the material, the greater is the tendency for the noise to be transferred to the vehicle occupant. All other factors being equal, a car made of a thinner material will be noisier than one made with the same material of a thicker gage.

Auto body material has come a long way from the 22-gage steel body that was typical of the early cars. With strides in material technology, automotive materials are now a mix of steel alloys—ranging from mild steel to ultra-high-strength steel, aluminum alloy strengthened with magnesium, and composite materials typically made of fiber-reinforced plastics. Though cost and ease of manufacture are major considerations in material selection, light-weight consideration has become equally important, particularly because of its effect on fuel efficiency.

The importance of material thickness to noise attenuation is in conflict with the light-weight goal that has become so important in modern vehicle design.

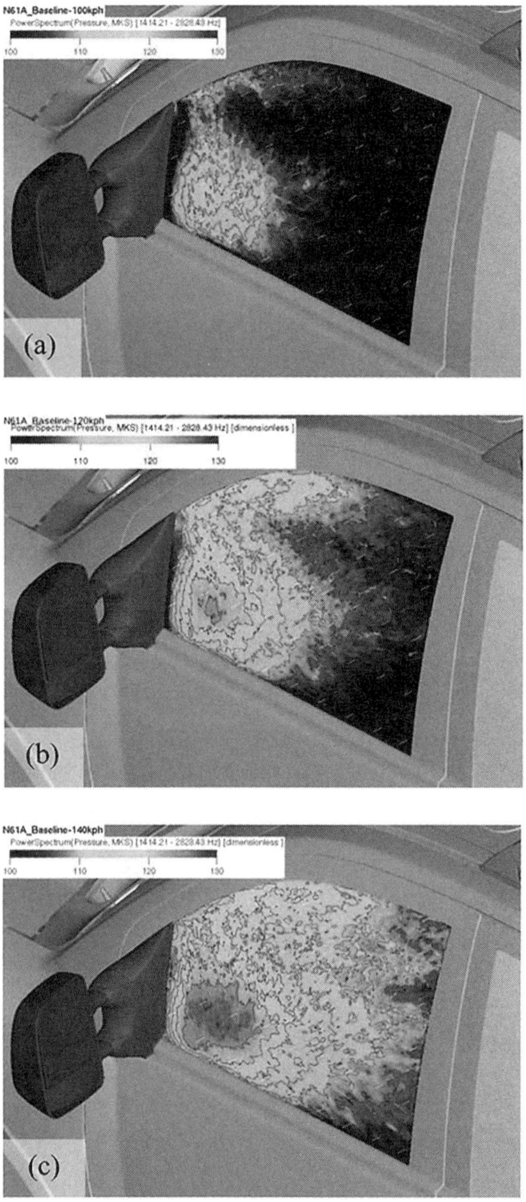

Fig. 2.3 Side glass noise: 2000 Hz octave band inlet source; vehicle speed (km/h) **(a)** 100; **(b)** 120; **(c)** 140 (reproduced with the permission of SAE International)

The compromise is a material with improved stiffness. The stiffer the material, the less it will flutter under windy or other noise-generating conditions. Sheet metals generally function as vehicle skin with stiffeners attached from the inside. Composite materials have inherent stiffness built into them and so generally do not need stiffeners to hold their shape.

The design of the vehicle itself plays an equally important roll in the level of noise transferred to the passenger. A good shock absorber system reduces bounce; well-oiled linkages and adequate stiffeners for holding the car skin firm to the chassis reduce rattle; and dynamically stable chassis tremendously reduces shake.

2.1.2 Windshield and Side Glasses

Windshield and side window buffeting create tremendous noise and vibration in a vehicle (Fig. 2.3). The pressure oscillations of buffeting are usually large and create a major discomfort for vehicle occupants. In addition, the A-pillar generates vortices, which compound the noise from the side glass.

2.1.3 Powertrain

Powertrain noise (or engine noise or underhood noise) reduction is particularly challenging because of the conflicting demands of lightweight materials and increased specific power in vehicle design. The source of this noise could be the connecting rods, poppet valve, crankshaft, rocker arm, engine cylinder block, or any combination thereof. It often manifests as rattle noise.

Powertrain noise can be detected with a stethoscope or with merely a well-trained ear. Improvements in techniques to detect powertrain noise include such acoustic holographic techniques as near-field acoustic holography (NAH) and spatial transformation of sound field (STSF). Dynamic balancing of the crankshaft substantially reduces engine vibration and the consequential powertrain noise. Statistical methods such as the design of experiments (DOE) have been successfully used to determine major noise transmission paths and approaches to refinement, which resulted in a substantial reduction of powertrain noise as well as a reduction in preproduction powertrain mounting development time [2-9]. Bounce, shake, and rattle (BSR) tests are also conducted on engine-on-the-mount to simulate the driving conditions the engine will be subjected to. The results are used to improve the underhood assembly for noise reduction, as well as for durability.

2.1.4 HVAC System

The noise caused by the HVAC ducts is strong. When the HVAC system is turned on, the noise intensity through the ducts can be of a broadband that is as high as 80 dB at a frequency of about 400 Hz. That is 10 dB more than the noise level in busy traffic and about the same as the noise level from a loud radio (Table 2.1), with a major contribution of the noise generation coming from the center duct. Because of the limited space between the HVAC and the outlet vents, the ventilation duct, a complex three-dimensional design, is too short to allow for full development and expansion of the air flow through it and the plenums. These sudden changes in duct dimensions and flow directions trigger flow separation, vortex shedding, and unwanted flow circulation, as well as other complicated turbulent flow structure inside the air ducts. These features contribute to increasing noise generation in the HVAC system. The higher the frequency, the higher is the pitch. The intensity and (to a lesser extent) the frequency have a direct bearing on loudness. A high-pitched and loud noise can have a range of effects from simply an irritation to causing passenger restlessness as the nerves involuntarily trigger in response to the increasing frequency of the generated noise.

2.1.5 Environment

Environmental noise consists of the noise due to the weather—wind, rain, and thunder—and the road surface condition. Depending on the texture of the road surface finish, the drive can be quiet and comfortable or very loud, even on a smooth and level road. The coarser the road surface finish, the greater is the tendency to experience a loud environmental noise, and this noise intensity increases exponentially with vehicle speed. This is also called the tire-road noise.

Wind noise is the dominant weather-based environmental noise. It is also the dominant noise for vehicle speeds above 30 mph (50 km/h), overwhelming the summation of tire noise, powertrain noise, and other ambient noise. Though wind noise can be generated as an exclusive ambient phenomenon, such as from a yell by a passerby on the street or the toll of a church bell, sources of wind noise are usually found in air-structure interaction and include the pillars (especially the A-pillars), side-view mirrors, side glass, doors, leaks around windshield and doors, roof racks, and exterior air flow that causes flutter of the vehicle skin, especially if the skin is made of thin-grade sheet metal.

2.1.6 Tire

Tire noise is two-part: the noise due to the tire geometry and texture and that due to the road. The latter is the tire-road noise referred to earlier, under environmental noise. A drainage pavement, that is, one constructed to be air and water permeable, is quieter than a more tightly packed pavement. The drainage pavement inherently has sound absorptive characteristics, which reduce tire noise generation and propagation. Also, open-graded pavement and rubberized asphalt concrete show improved tire-road noise over a standard Portland cement concrete (PCC) surface of more-or-less the same material aggregate [2-10].

A wider tire, although providing better vehicle stability, is noisier. It is also more expensive in its direct cost and in the cost of vehicle operation. Because it is heavier, a greater brake power, hence more fuel consumption, is required to overcome the added weight that it incurs.

At low speeds, tire noise dominates the wind noise, but as speed increases, the wind noise increasingly becomes the major noise in ground vehicles.

2.2 Types of Aerodynamic Noise

There are different ways noise can be classified, but for our purpose, which is aerodynamic noise, we will classify noise into four types: attached flow noise, separated flow noise, cavity noise, and aspiration noise.

2.2.1 Attached Flow Noise

This is noise due to wind-vehicle body interaction. The constant compression and rarefaction of air molecules in a cross wind impart a pulsating pressure on the windows and on the skin of a car, interacting with the drag that the vehicle is already experiencing. If the flow stream along the vehicle's longitudinal axis is strong enough, it is maintained in spite of the lateral air pulsation; the passenger only hears a little rattle of the window and a little car-body noise even in a strong cross-wind (Fig. 2.4). There is no relationship between the intensity of the noise and the car's aerodynamic drag. The attached flow may be laminar or turbulent, depending on the car velocity and on such properties as the car-body smoothness and the shape. Attached flow noise can further be classified as monopole, dipole, and quadrupole.

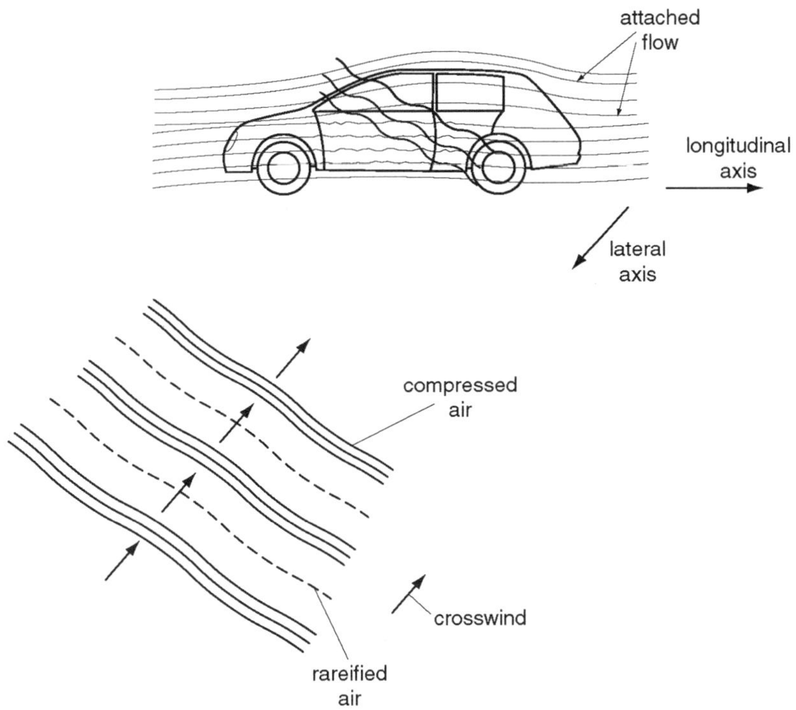

Fig. 2.4 Attached flow noise

A monopole sound is radiated equally from its source into all directions. The monopole source creates sound waves by alternately introducing fluid into and removing fluid from the surrounding area. The sound intensity of monopole sounds is of the fourth order of the velocity (v4) (Fig. 2.5a). A contained loud speaker at low frequencies is an example of a monopole source.

A dipole sound is radiated by two monopole sources having equal strength but opposite phase and separated by a distance, Δd, which is small relative to the wavelength of sound ($\Delta d < \lambda$). In a dipole system, one source expands and the other contracts simultaneously, causing the air near both sources to undulate as it produces sound (Fig. 2.5(b)). The sound intensity of dipole sounds is of the sixth order of the velocity (v6). A loudspeaker that is not contained is an example of dipole source.

A quadrupole sound comes from two opposite dipoles. If the two opposite dipoles lie along the same line, they make a linear quadrupole (Fig. 2.5(c)). Otherwise, they are called lateral quadrupole. In a lateral quadrupole the four monopoles are arranged at alternating phase at the corners of square (Fig. 2.5(d)). The sound intensity of quadrupole sounds is of the eighth order of the velocity (v8).

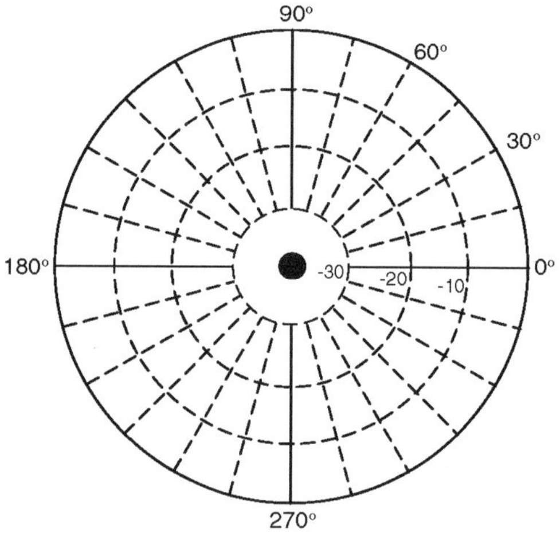

Fig. 2.5(a) monopole

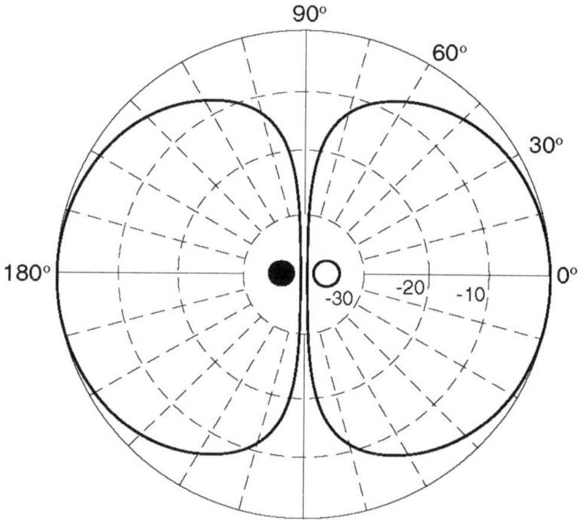

Fig. 2.5(b) dipole

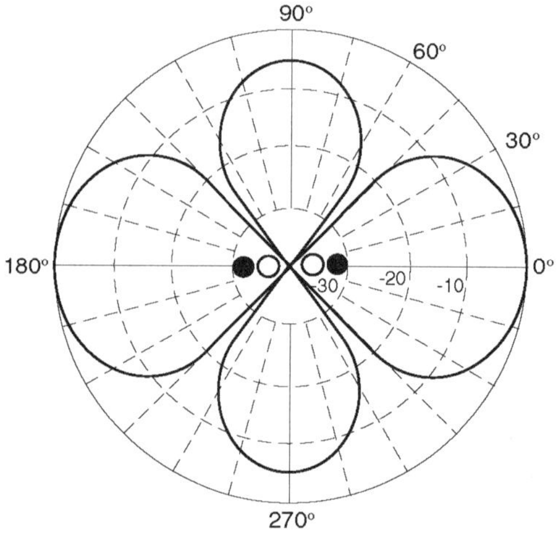

Fig. 2.5(c) linear quadrupole

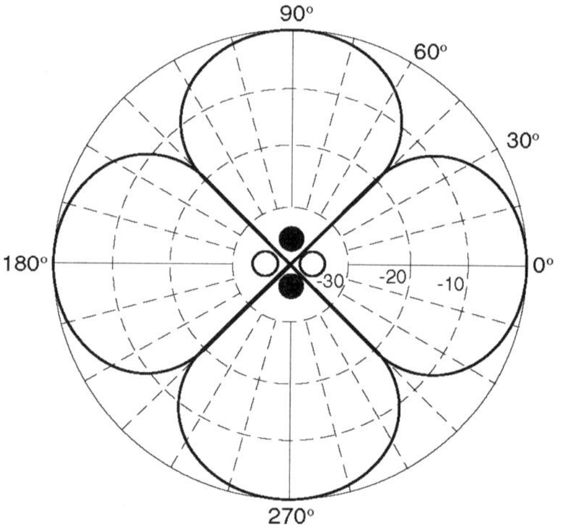

Fig. 2.5(d) lateral quadrupole

Fig. 2.5 Classification of noise sources.

Although common in the cross-wind direction, attached flow noise can occur in a head wind or a rear wind, hitting the front windshield and the rear windshield, respectively.

2.2.2 Separated Flow Noise

In separated flow noise, the lateral air pulsation interacts with turbulent, separated flow, and may increase the turbulence level of the flow. Strouhal number, *St*, is a nondimensional term used to quantify the intensity of noise and to compare the extent of oscillating, unsteady flows in various designs. It is defined as

$$St = \frac{nd}{v_i}$$

where

n is the frequency of vortex separation from the body (or the oscillation frequency);

d is the hydraulic diameter;

v_i is the flow velocity.

The Strouhal number represents a measure of the ratio of inertial forces due to the unsteadiness of the flow, or local acceleration to the inertial forces due to changes in velocity from one point to another in the flow field. The trailing vortex of vehicles is an example of the application of Strouhal number. When the frequency of the vortex separation from the body is in the audible range (20 Hz to 20 KHz), the pressure fluctuations within the unsteady flow translate to noise. The greater the fluctuation, the higher the Strouhal number, up until about 0.2, where it usually levels off.

2.2.3 Cavity Noise

Cavity noise is that generated when head-on wind hits crevices and external gaps. Convertibles, sunroofs, and roof-racks are examples of sources of cavity noise. Cavity noise is also generated when the car crosses an expansion joint.

2.2.4 Wind Aspiration Noise

Aspiration noise is air leaking into or out of the car through the door seals.

2.3 Locating Noise Source

To abate noise in a vehicle, the sources of noise must first be located. This is still done by a prolonged manual process. The following steps are taken to isolate the source of noise in a car passenger compartment:

- wind up all windows
- tape probable wind sources (e.g., door gaps)
- remove interference components (e.g., windshield wipers)
- measure interior sound level (intensity and frequency)—this is reference level
- expose area of vehicle to wind

- measure and record increase in interior wind noise
- locate a crack (i.e., a source of noise)
- tape off that crack
- measure and record new noise level
- locate another crack, tape it off, and measure new level

The steps are repeated until all the leaks are identified. The leaks are usually found to be the result of ineffective sealing. A poor quality seal or a poorly placed good quality seal will both allow noise through.

2.4 Noise Measurement

In-cabin vehicle noise is usually taken at a speed range of 50 mph (80 km/h) to 70 mph (113 km/h) and are commonly measured in dB SPL and presented in max/min format for a given speed and road condition.

Qualitative sound variation can be done with a trained human ear. This may be a valuable first pass, prior to instrumentation for noise measurement.

2.4.1 Microphones

The microphone is a common noise measurement device. The placement of the microphone is critical to the correctness of the obtained result. Some microphones are used for near-field (i.e., in-cabin) acoustic measurement, and others are determined to be best for free-field placement. It is crucial that the microphone is fully within the noise cone to capture a complete, rather than a partial, noise level reading. The technique of NAH has been found to be dependable in locating noise sources and in identifying flow paths for both interior and exterior fields.

2.4.2 Acoustic Dummies

Much like the dummies used in vehicle crash tests, acoustic dummies are rigged with data recorders and placed (usually) in the back seat of a car. Sometimes, especially with luxury cars, the acoustic dummy test is performed with two dummies placed at the two corners of the back seat, and the respective recordings are synchronized.

2.4.3 Sound Level Meters

A sound level meter is in essence an advance microphone. Depending on the quality and variety, it could consist of a microphone, an amplifier, an output meter, and frequency weighting networks in one package.

2.4.4 Accelerometers and Dynamometers

An accelerometer is a refining tool for low-level noise measurement. It is secondary to such devices as the microphone or the acoustic dummy. A dynamometer is typically used for measuring engine noise and vibration. Other forms of noise recording devices include digital recorders and acoustic circuits.

2.5 Material-Based Noise Abatement

The goal is not always the entire elimination of noise. Besides, noise cannot be entirely eliminated, even in high-end cars. Noise reduction to an acceptable level is the primary interest in noise abatement. There are materials to employ as well as design modifications that can be implemented for vehicle noise reduction.

2.5.1 Acoustic Foam

Acoustic foam (Fig. 2.6), with its high absorption-to-thickness ratio, does serve as a worthwhile noise abatement device. Placed in the gut of the doors, it absorbs most of the far-field noise. It is important to ensure that the foam has effective resistance properties for air flow, heat, humidity, flame, and chemicals. It should also have good damping properties over a broad temperature range and frequency range, especially when it is used as a hood liner, functioning particularly to effect adequate sealing of joints between the hood and the radiator grille.

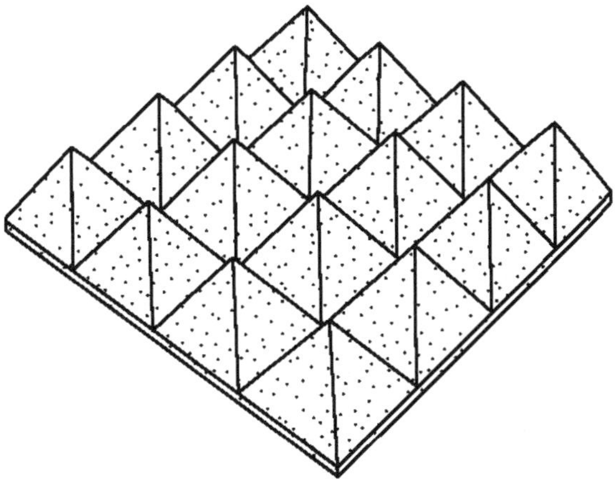

Fig. 2.6 Acoustic foam

2.5.2 Acoustic Barrier

The foam is a sound-deadening material. An alternative would be to use a sound-deflecting material, such as sound resistant linoleum lining of the inner skin, or thick sheet metal. The use of thick sheet metal has its drawbacks, and so it is usually not a practiced approach, especially in the age of design for maximum output, power-to-fuel, output-power-to-weight ratio. The optimum approach is to introduce a damping package at various points of the body between the not-so-thick sheet metal and the inner upholstery. This substantially reduces the low-frequency noise.

2.6 Design-Based Noise Abatement

There are a number of design modifications that can be made to vehicle design to reduce noise (Fig. 2.7). These modifications are more important in car design (except for race cars) than in trucks and trailers. Luxury buses would also benefit from such design modifications. Listed below are some of the modifications that result in the desired reduction of the noise level transmitted to the vehicle occupant.

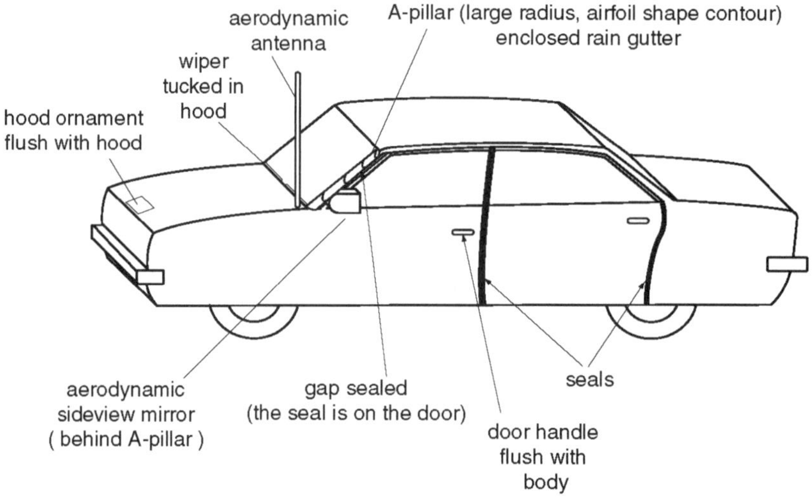

Fig. 2.7 Design for noise reduction

2.6.1 A-pillar

The A-pillar is at between a quarter and a third the length of a car—it is at about an eight the length of a tractor trailer. Here are some steps to ameliorate the noise tendency due to the A-pillar:

- Design with large radius
- Contour like an airfoil to reduce the turbulence intensity of separated flows
- Avoid exposed rain gutters
- Seal A-pillar gap on doors having full-frame window design

2.6.2 Side-View Mirrors

Move mirrors behind the A-pillars to reduce separation and turbulence level at the pillars:

- Position the mirror such that its base is just above the beltline
- Round the mirror housing (this both reduces noise and decreases aerodynamic drag)
- Project the mirrors from the body surface (this nearly eliminates the interference drag, improves aerodynamic drag, and reduces wind noise on the side glass surface)
- Avoid gaps and drain holes in mirror housing (design housing with a downward slope for self-draining)
- Avoid exterior screws in mirror housing (exterior screws can trigger turbulence)

2.6.3 Windows and Sunroof

- Form the sunroof into a profile that geometrically blends with the rest of the car top, effectively sealing the glass-metal interface
- For the tilting sunroof, implement an effect wind deflector on the trailing edge of the sunroof
- For the sliding sunroof, implement an effective wind deflector on the leading edge of the sunroof
- Minimize side glass offset to pillars
- Use a flush surface for side glass-pillar interface to ensure low drag
- Use thicker window glasses to reduce noise transmission through glass

2.6.4 Doors and Door Handle

- Integrate the front portion of the door handle such that it is an aerodynamic continuation of the body
- Flat airplane type door handle is ideal (it flushes with the car body)
- Provide durable sealing of the door handle
- Improve the rigidity of the doors

2.6.5 Windshield Wiper

Contain the wiper under the hood when not in use so that it does not trigger turbulence. By design, the location of the windshield wiper is already favorable for the generation of local flow separation and turbulence. This tendency is made worse by the exposure of windshield wipers. It is important, both for noise reduction and for drag reduction, to contain the windshield wiper under the hood.

2.6.6 Gaps

- Use auxiliary seals for gaps between vehicle body and doors
- Use primary seals to prevent wind noise transmission into passenger compartment
- Seal drain holes where feasible
- Use quality adhesive for seals so that cracks do not develop within the adhesive, over time

2.6.7 Hood Ornament and Radio Antenna

Hood ornaments that are flush with the hood of the vehicle are by far preferred to those that stick out, generating drag and creating noise. Even when aerodynamically carved, hood ornaments still add to vehicle drag, but not as much to noise. It is preferable to eliminate antennas, or a telescopic design is recommended so that it is collapsed into the car when not needed. Wrapping a helical strip around a fixed antenna helps, to some extent, to reduce the noise and the aerodynamic drag.

2.7 Wind Tunnel Noise

In chapter 3, wind tunnel testing is introduced. The engines driving the propeller blades are usually in the order of 200 hp (150 kW), driving air at speeds of about 125 mph (200 km/h). A lot of noise is generated in such an operation, even with a low-noise fan. It is important to minimize wind tunnel noise because the noise factor can influence the result of wind tunnel tests, especially if it is not a comparative test. Wind tunnel noise becomes significant at frequencies above 500 Hz.

Fan tip speed and fan efficiency:

- Select a lower tip speed fan design for a given efficiency

Test section plenum walls:

- Reinforcement is important, especially for pressure tunnels
- Design for access to test section under pressure

Diffuser and contraction cone (Fig. 2.8)

- The included divergence angle of diffuser walls < 10°
- Install silencers in circuit return path
- Aerodynamically efficient corner vanes leading to the contraction cone
- Sound absorbing materials, such as foam and/or porous concrete wall on the outside surface of the contraction cone

Silencers may also be installed in the return path of the circuit.

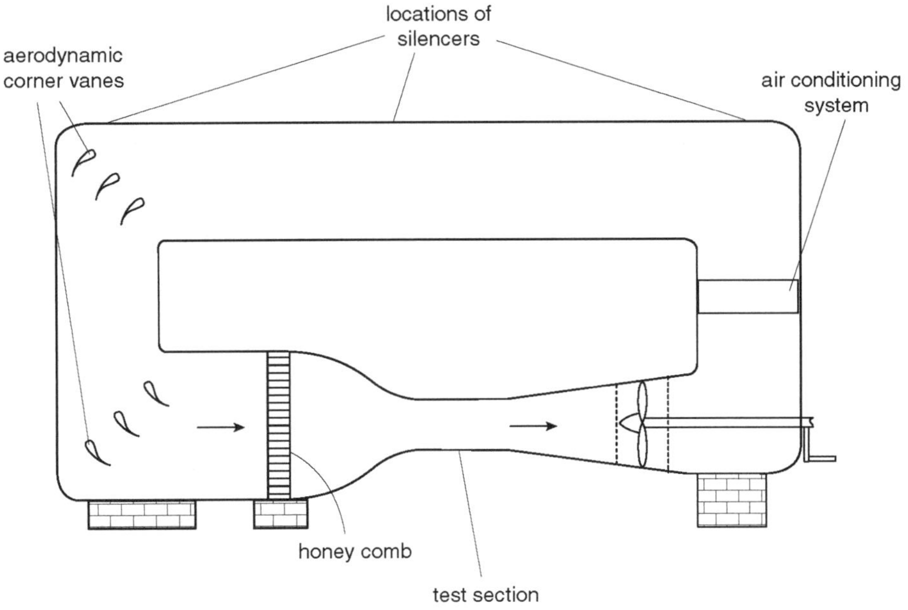

Fig. 2.8 Wind tunnel noise reduction

2.8 Vehicle Soiling

Vehicle soiling has only recently come to the forefront of considerations in vehicle design. Vehicle soiling may be defined as the clouding effect of fluid particles or fluid-entrained particles on the external surface of a vehicle (Fig. 2.9).

Fig. 2.9 Vehicle soiling (Photo courtesy of Jaguar Land Rover/Exa Corporation. Used with permission.)

Soiling of road vehicles is effected by a combination of splash and spray. Splash refers to the large droplets raised by the tires. It typically follows a ballistic trajectory, settling either on the vehicle surface or on the ground. Spray, however, is made up of tiny droplets that remain suspended in the atmosphere for an extended period, producing fog clouds and inhibiting visibility. It is not so much the splash as it is the spray that causes at least inconvenience and serious dangers of operator visibility impairment and harm to pedestrians. Rain, snow, dust, and mud are all vehicle soiling agents. Greater attention is understandably placed on mud soiling than on the other forms, but the mode and effect of soiling are the same in all its forms. From an aerodynamic perspective, there are two types of soiling, namely streamlined flow soiling and turbulent flow soiling. Dirt deposit rate is considerably more in the soiling over turbulent flow regions than it is in the streamlined flow regions. The term dirt as used in this context refers to the soil (i.e., rain, snow, dust, mud). Soil is commonly referred to as dirt because in most of the studies on soiling, attention has been placed exclusively on the soiling from the ground.

Flow separation is one of the instigators of turbulent flow; therefore, considerable soiling is to be expected in the parts, particularly in the lower parts (i.e., below the mid-horizontal plane), of the vehicle profile where flow separation is imminent. The soiling above the mid-horizontal plane will generally be referred to as glass soiling, and that which takes place below mid-horizontal plane is body soiling.[1]

2.8.1 Front Soiling

Because of the teardrop leading-edge design typical of a car's fore section, flow separation, hence large-scale soiling, is uncommon over the leading edge of a car. There is still some soiling that usually comes from the vehicle ahead or from the elements rather than being self-provoked. An example of front soiling is the dirt on the headlights and on the front-side indicators. Another is the rain and snow accumulation on the windshield.

It is of paramount importance for the driver to be able to see clearly through the windshield. The singularly most effective attenuator of windshield soiling is the windshield wiper. It is therefore important to keep the wiper blades in good condition and to ensure that the wipers effectively grip the windscreen.

Additional reduction in front soiling can be obtained by reducing the gradient of the windshield, and by increasing the hood camber and the lateral curvature of the windscreen. The same design has an aerodynamic benefit by reducing the drag contribution from the windshield and the hood (see Section 6.1).

2.8.2 Rear soiling

The opposite is the case for the rear of a vehicle (Fig. 2.10). Here, soiling is typically self-provoked.

[1] The terms are not exact because the vehicle center of gravity is slightly below the lower end of the glass placement. But the approximation is qualitatively adequate.

Fig. 2.10 Rear soiling: **(a)** before soiling and **(b)** after soiling (reproduced with the permission of SAE International)

As the vehicle travels over the road, dirt is usually raised from the road surface by turbulence that, due to air mixing within the wake, is deposited onto the rear surface—rear window, trunk, rear bumper—of a vehicle (and a residue extends toward the trailing vehicle, causing front soiling in the trailing vehicle). Depending on the weather, soiling due to the flow over the roof can become as aerodynamically significant as that from the vehicle's rear tires and underside, with its effect increasing as the vehicle speed increases. It has become a common practice to extend the rear edge of the roof of vans and station wagons to increase the flow attachment to the vehicle, thereby reducing the drag coefficient. It also improves the vehicle's stability. But this same aerodynamically beneficial vehicle addition may be unfavorable when evaluated for soiling. The flow from the attachment leaves the vehicle, allowing for the rising flow from the bottom to swirl its way up farther on the vehicle rear (Fig. 2.11). This flow from below carries with it the elements on the ground. The result is the soiling of the vehicle rear farther up than would be the case without the rear edge extension. A greater attention is therefore paid to the soiling from the vehicle's lower plane flow because it more readily inhibits rear visibility than the relatively clean flow from the vehicle roof.

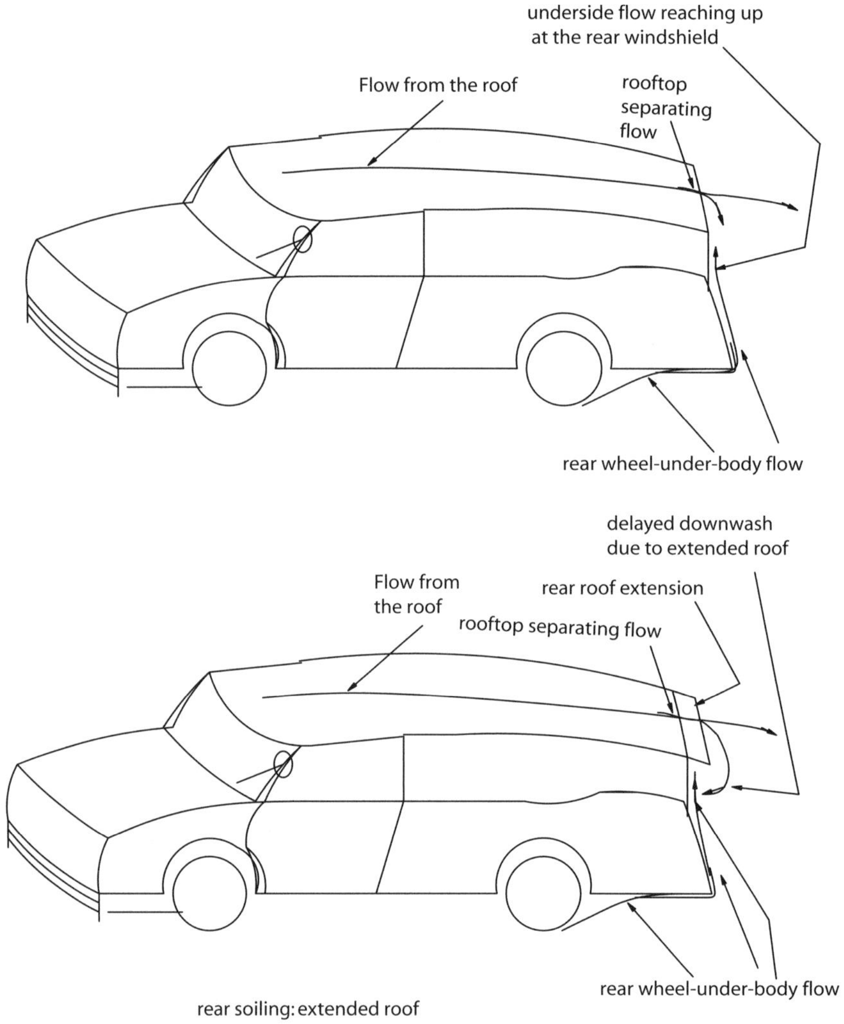

Fig. 2.11 Rear soiling with and without extended roof

Rear window soiling is more pronounced in a station wagon than in a sedan, with a hatchback and a fastback, respectively, showing decreasing rear soiling:

Highest rear soiling: station wagon > hatchback > fastback > sedan: Lowest rear soiling.

Let's consider the dynamics responsible for the difference in rear soiling between, say, the two extremes: station wagon and sedan (Fig. 2.12).

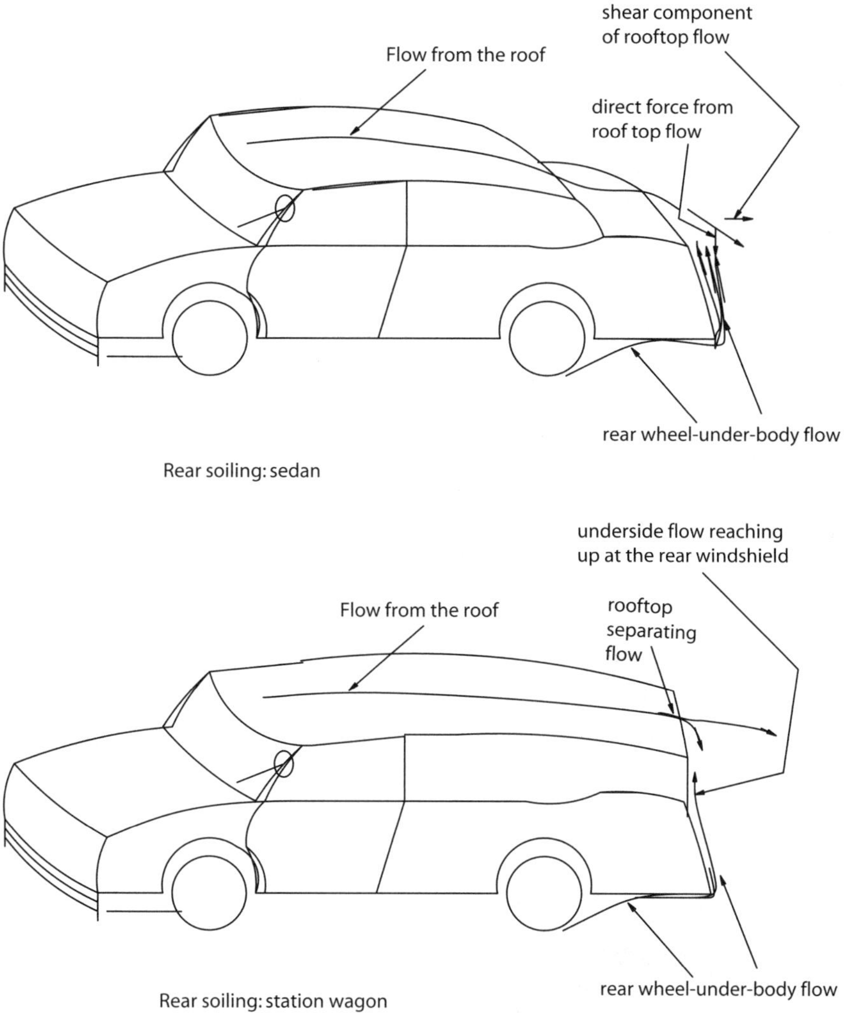

Fig. 2.12 Soiling comparison: sedan versus station wagon

The flow from the roof of the sedan follows the rear window as it inclines toward the bumper. The swirling flow from the rear wheel combines with that from the underside and turns upward. But it is met by the downward and rearward flow, which started from the roof. This downward flow has both a direct force and a shear force. The direct force acts against the upward trend of the flow from under, and the shear force reduces the upward momentum of the same flow (coming from under). The result is a rear soiling that is limited to the rear lower height of the trunk.

In the case of the station wagon, most of the fluid from the roof separates rather abruptly at the rear window. Some of the separated reverses and hits the rear window. This action reduces the downward force of the part that flows down the rear window. Also, the station wagon does not have the shearing component present in the sedan as described earlier. The result of the two disadvantages is that the flow from the vehicle lower region is able to reach up higher in the station wagon than in the sedan, and hence a greater soiling of the rear window.

One of the efforts to improve the rear soiling is the introduction of a guide vane at the rear roof of the vehicle. The guide vane helps to collect the soil and direct it downward on the rear edges rather than all over the rear face. This, however, may cause an increase in the drag coefficient at the rear window.

Another consideration for improvement is to design the rear lower section so as to avoid stirring up road surface debris by the wake turbulence and to deflect the minimal debris to the road surface, as in the use of mud guards. The vehicle designer should also arrange the air flow to induce a wash on the rear window and to prevent the settling of dirt particles.

2.8.3 Side Soiling

Side soiling comes from three sources: the front wheel, the rear wheel, and external sources.

Front wheel. The flow from the rotating front wheel impinges on the wheel housing and is deflected downward and outward of the vehicle as it is gathered with the relatively high-speed main flow of the vehicle. Any dirt carried in the rotating flow is therefore transferred to the side of the vehicle as part of the shear flow on the vehicle lower skin and to the surroundings (Figs. 2.13 and 2.14).

(a) Intensity of body side deposition from front wheel

(b) Splash and drip from front wheel housing and rocker panel

Fig. 2.13 Side soiling (reproduced with the permission of SAE International)

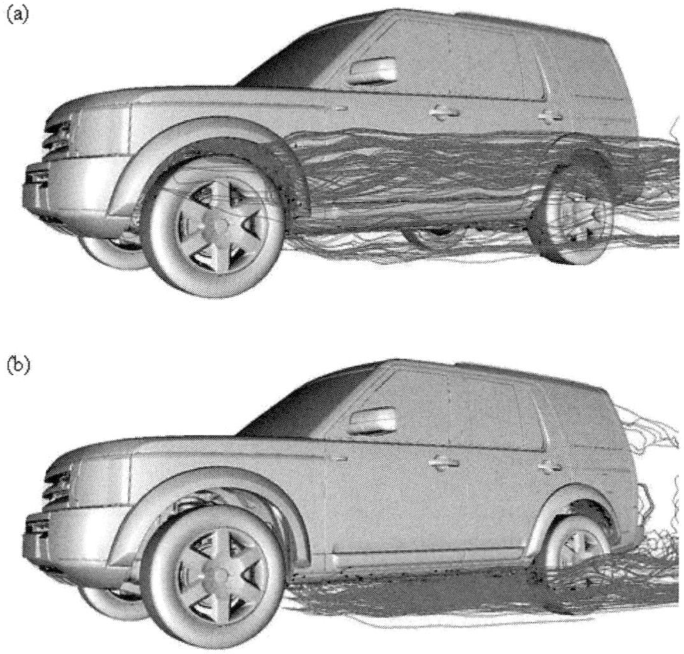

Fig. 2.14 Particle traces from dripping particles. **(a)** Low-velocity particles released from wheel housing; **(b)** Low-velocity particles released from rocker panel (reproduced with the permission of SAE International)

Rear wheel. The mode of dirt dispersion from the rear wheel is similar to the case of the front wheel, but the effect is not as pronounced on the vehicle side as that from the front wheel mainly because of the location of the rear wheel in the vehicle architecture. To a lesser degree, the flow from the rotating rear wheel impinges on the wheel housing, the remaining portion shooting to the rear of the vehicle with flow components in the longitudinal, lateral, and vertical directions. This dirt-included flow combines with the rear underbody flow, thus gaining momentum, lifts up additional dirt from the ground, and entrains in the main flow around the vehicle with the mixture forming the trailing vortex. The rear wheel contribution to side soiling is thus mainly on the side of the vehicle behind the rear wheel (Fig. 2.15).

External sources. External sources include side wind and other passing vehicles. (Egg pelting is a controllable external source which will not be considered because of its rarity!) Gusts flowing in the vehicle's lateral direction can hit the vehicle with such a direct compressive force that can cause the included dirt to "glue" onto the side of the vehicle. Sometimes, the wind speed is higher than the vehicle speed. Apart from the maneuvering challenge that accompanies such situations, especially when it is a cross wind, there is also a "trapping down" of the vehicle, and this subjects the vehicle's side to all that the side wind can bring upon it. Passing vehicles are, in essence, nearby sources of wet side

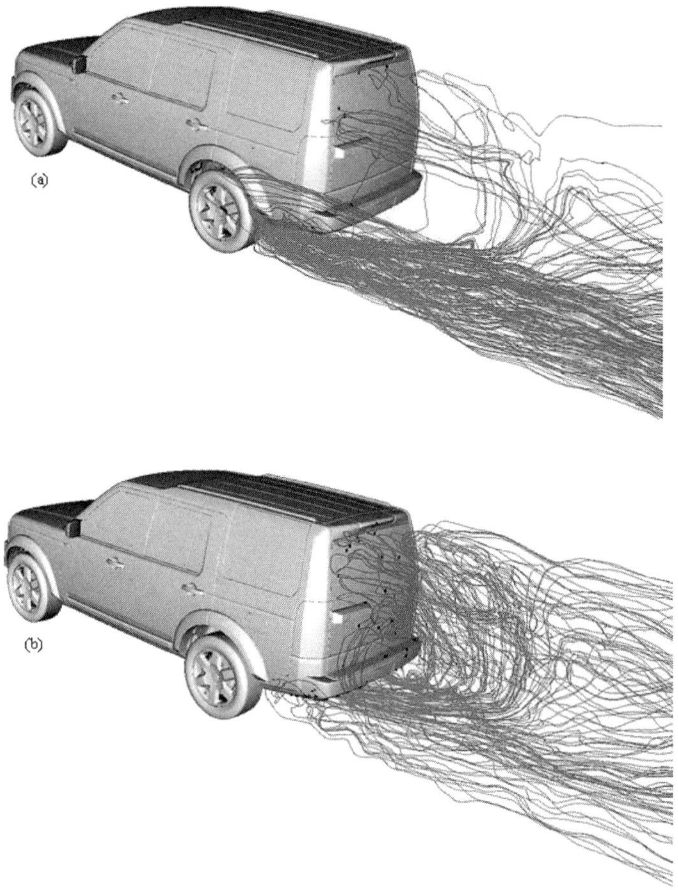

Fig. 2.15 Particle traces from dripping particles. **(a)** From wheel house; **(b)** From rear underbody near rear bumper (reproduced with the permission of SAE International)

wind. If a passing vehicle goes through a puddle, its wheels can raise the area of wet fluid generated to cover the entire side of a nearby vehicle, especially if the affected vehicle is smaller than the inflicting vehicle. (This is one reason to wind up the windows when driving through a wet road.)

2.8.4 Engine Intake Soiling

A common consequence of engine intake soiling is air filter clogging. Engine intake is not particularly designed with consideration for soiling. It may become a design consideration as areas of improvement in vehicle performance become narrower. Mud and heavy snow trapped in the engine compartment add to vehicle weight; this reduces fuel efficiency. The design of the lower underhood can be improved so that it does not catch and keep the snow and mud that tend to hide therein.

In whichever form it occurs, soiling is, in the least, inconvenient; more seriously, it can be dangerous. Soiling defaces, albeit temporarily, the appearance of a vehicle. It limits driver visibility when it is deposited on the windscreen. Rear soiling, apart from the self-soiling effect, inhibits the vision of trailing vehicles, especially if the trailing vehicle is relatively smaller in size. Soiling significantly contributes to the intensity of wind noise experienced by vehicle occupant, especially at high vehicle speeds. Passengers may soil their clothes as they get in and out of the vehicle. An equally important consequence of soiling is the covering up of road signs and road structures such as partition lines and guard-rails. The importance of soiling elimination or, at least, minimization should not be underestimated.

The shape of a vehicle will determine the flow pattern around the vehicle and the extent to which soiling affects the vehicle. Unfortunately, it is not realistic to provide a general aerodynamic equation or an all-encompassing design to effect the near-elimination of soiling. The overall aerodynamic adequacy of a vehicle is generally evaluated qualitatively using flow visualization techniques, the most common of which is smoke generation. Tests must be performed for each vehicle profile to find the design envelope that both minimizes stirring up of dirt and maximizes the air flow induced wash on the vehicle in general, and on the windows in particular. Yet such a design should not take away from the already accomplished vehicle aerodynamic optimization.

Summary

Noise creates discomfort for the driver and other vehicle occupants. It can also cause distraction for the driver. Wind noise become significant to the vehicle occupant at speeds above 50 mph (80 km/h). Noise intensity is defined as the noise energy rate (power) per unit area and is expressed in watts per meter squared. Noise level is a logarithmic representation of the intensity and is expressed in decibels. The relationship between noise level (decibel) and noise intensity (watt per meter squared) is

$$\beta = 10\log_{10}(I/I_o)$$

where

$I_o = 10^{-12}$ W/m^2, the threshold of hearing at a frequency of 1000 Hz.

The primary sources of aerodynamic noise are the vehicle body and the side glasses.

Vehicle noise can be reduced by methods that include the use of noise abatement materials and strategic component design for noise reduction.

Vehicle soiling is caused by a combination of splash and spray. Although both contribute to passenger discomfort, spray is the primary cause of driver visibility impairment.

Key Terms

Noise: Sound intensity that causes discomfort. (Although there are low-level noises, noise is usually associated with high intensity.)

Soiling: Clouding effect of fluid particles or fluid-entrained particles on the external surface of a vehicle. (Soiling of road vehicle is effected by a combination of splash and spray.)

Review Questions

1. Name three main sources of noise in ground vehicles
2. Describe aerodynamic noise
3. Why is aerodynamic noise the dominant noise mode at high speeds?
4. The two mechanisms of vehicle soiling are splash and spray. What is the difference between them? Which is probably of a greater concern from safety standpoint? Explain.

Application Question

Higher strength steel is being evaluated for vehicle bodies and structure. What do you think the effect will be on vehicle noise? Will it contribute to noise reduction or will it be detrimental to the goal of vehicle body noise reduction? Explain.

Exercises

1. How many times more intense than threshold is a sound level of 83 dB?
2. How many decibels higher is a sound of intensity 10^{-3} W/m^2 than one of intensity 10^{-10} W/m^2?
3. Which is more painful to the human ear, the noise from a busy traffic with restless drivers persistently blowing the horn or the noise from inside a subway train traveling on a yet-to-be upgraded track? Explain.

Projects

1. Some tires ("lugged" tires) are very noisy. How would you control the noise level of such tires?
2. Design a wiper blade that has a good "grip" of the windshield and minimal noise generation in operation.
3. Compare the noise level in a composite enclosure with that in a sheet metal enclosure. What are the advantages and disadvantages of each? Quantify an optimum design, given the two materials.

References

1. Lighthill, M. J., "On Sound Generated Aerodynamically: Turbulence as a Source of Sound," *Proc. R. Soc. Lond.* A 222, 1954, pp. 1–32.
2. Hamilton, M. F., and Morfey, C. L., *Model Equations, Nonlinear Acoustics,* Hamilton, M. F., and Blackstock, D. T., eds., Academic Press, 1998.
3. Landau, L. D., and Lifshitz, E. M., "Fluid Mechanics," in *Course of Theoretical Physics,* 2nd ed., Vol. 6, Butterworth-Heinemann, 1987.
4. Naugolnykh, K., and Ostrovsky, L., "Nonlinear Wave Processes in Acoustics," *Cambridge Texts in Applied Mathematics,* Vol. 9, Cambridge University Press, 1998.
5. Prout, J., *Acoustics for You,* Krieger Publishing Company, 1990.
6. Morse, P. M., and Ingard, K. U., *Theoretical Acoustics,* Princeton University Press, 1986.
7. Nelkon, M., and Parker, P., *Advanced Level Physics,* 3rd ed., with SI Units, Heinemann Educational Books Ltd. 1970, Chpts. 25–27.
8. Blumrich, R., "New Developments in Numerical Vehicle Aeroacoustics," 7th FKFS Conference Progress in Vehicle Aerodynamics and Thermal Management, Stuttgart, July 2009.
9. Wang, X., and Ren, H., "Vehicle Power-train Noise Refinement Using the Taguchi Method," *International Journal of Vehicle Noise and Vibration,* 2007, Vol. 3, No. 3, pp. 286–301.
10. Donovan, P. R., "Comparative Measurements of Tire/Pavement Noise in Europe and the United States–A Summary of the NITE Study," California Department of Transportation.

Chapter 3
Experimental Aerodynamics for Ground Vehicles

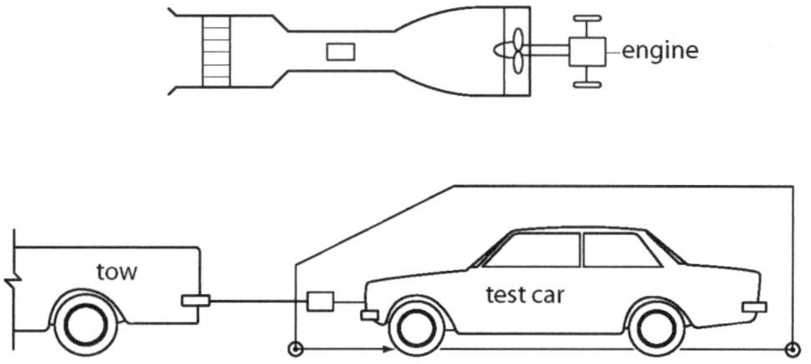

Experiment and theory are much more interactive than they are sequential, at least until a theory has become well established, thanks to the validating results from experiment. The success of computational physics lies heavily on reliable experimental results. Experiments are performed either in real-life road and vehicle conditions or in simulated conditions. While real road condition experimentation is preferred, it is unrealistic both in cost and in the ready availability of desired conditions. Most experimental work is therefore performed in a wind tunnel, usually with a geometrically scaled-down model of the test object [3-1, 3-2, 3-3, 3-4, 3-5, 3-6].

3.1 Wind Tunnel

Once the mission specification has been established for a ground vehicle design, one of the early decisions of the vehicle aerodynamic group is to determine the type of wind tunnel testing to perform and to choose the appropriate wind tunnel and scale model for evaluating the concept design.

3.1.1 Principles of Operation

A wind tunnel is a controlled environment where air speed, flow direction, temperature, and other variables are not influenced by outdoor conditions. Test conditions are well controlled; the test model is fixed in its stationary position, whereas air movement through the test section is controlled (Fig. 3.1).

3.1.2 Wind Tunnel Classification

Wind tunnels are classified by several methods. They include classification by architecture, speed, air pressure, and size.

3.1.2.1 Classification by Basic Architecture

A wind tunnel can be open circuit or closed circuit. In an open circuit wind tunnel (Fig. 3.1), air flows straight from the surrounding environment into the honeycomb, which straightens the flow. The air travels through the contraction cone, which, consistent with Bernoulli's principle, speeds up the flow to give it the highest speed at the test section. Thereafter, the flow slows down at the diffuser and exits into the environment. The contraction cone is critical to the maintenance of a streamlined flow through the test section. If the contraction ratio is too high, a deflected, turbulent flow will result in the test section [3-7].

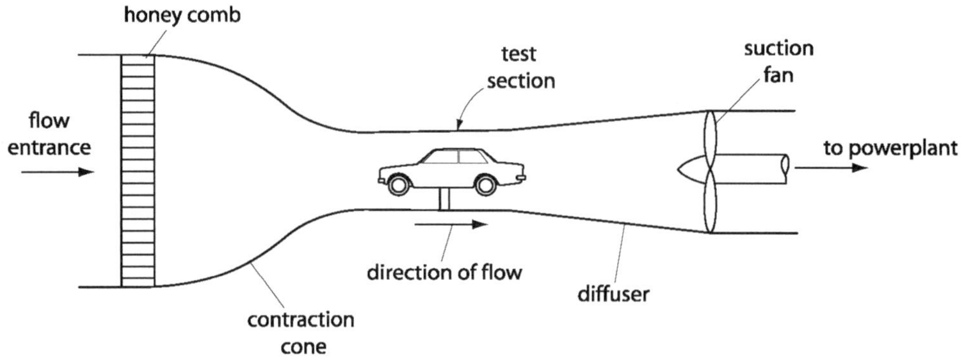

Fig. 3.1 Open-circuit wind tunnel

In the closed-circuit wind tunnel, air is contained and recycled in the system (Fig. 3.2). Air is pulled in by the fan, cycled through the return duct until it reaches the settling chamber, where it is straightened before it proceeds through the contraction cone to the test section. The same air is recirculated within the tunnel for the duration of the test. This preserves the test from the effect of any variation in the ambient air.

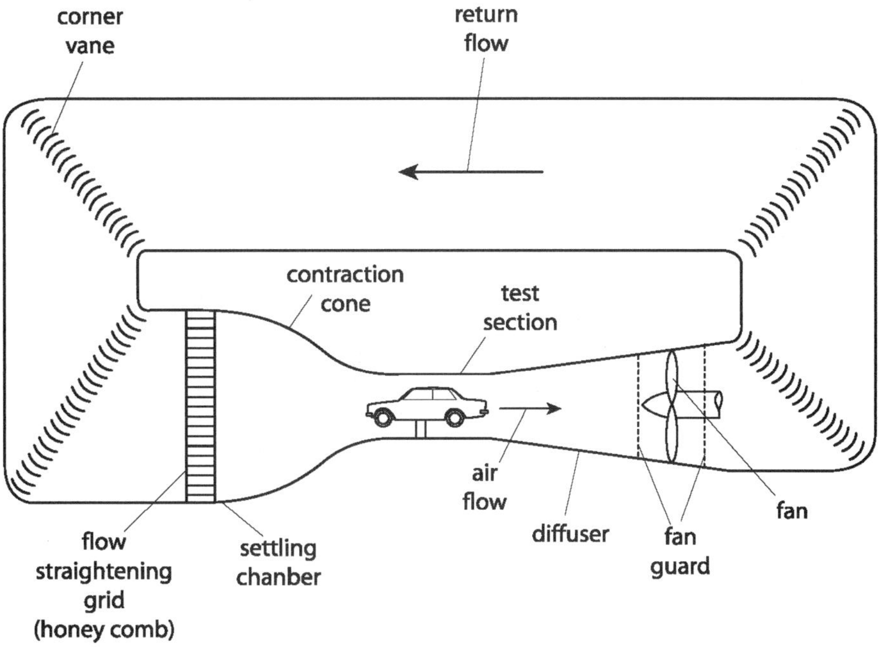

Fig. 3.2 Closed-circuit wind tunnel

In a further architectural classification, the wind tunnel is described by the type of test section—open or closed. We can have, for example, a closed-circuit open test section wind tunnel. The loop and test-section combination gives the following possible compound classifications of a wind tunnel:

Open circuit, open test section

Open circuit, closed test section

Closed circuit, open test section

Closed circuit, closed test section

In closed test section wind tunnels, the walls are sometimes slotted to allow for air exchange with the outside. After an extended, continuous use, the tunnel air heats up, causing a change in the ambient condition. To avoid this, slots in the tunnel wall allow the infusion of fresh air and the bleeding of the now heating-up tunnel air to maintain a more or less consistent temperature and pressure in the test section [3-8, 3-9, 3-10, 3-11, 3-12].

The open test section can be fully open or partially open—open ceiling, open side, open floor.

The test object may be on a fixed floor in the test section or on a moving belt floor. A moving belt floor is introduced to minimize the build-up of a boundary layer in the test section.

3.1.2.2 Classification by Speed

There are four types of speed-rated wind tunnels: subsonic, transonic, supersonic, and hypersonic [3-13, 3-14, 3-15]. The common type is the subsonic wind tunnel.

A wind tunnel is defined as subsonic when its maximum speed is below Mach 0.8 or 609 mph (980 km/h). The subsonic wind tunnels are of greater interest to the ground vehicle aerodynamicist. Subsonic wind tunnels are sometimes further classified into low-speed subsonic (<Mach 0.4) and high-speed subsonic (>Mach 0.4).

The transonic range is Mach 0.8 to 1.2 (i.e., 609 to 913 mph—980 to 1470 km/h). Transonic wind tunnels are designed on the same principles as the subsonic wind tunnels in that the highest speed is reached in the test section. Slotted walls are usually included in the design of transonic wind tunnels; this is to reduce shock reflection from the walls of the test section. In addition to the Reynolds number, which is important for all speed ranges, Mach number starts to be important for transonic flows due to the development of shock waves.

The supersonic range is Mach 1.2 to 5 (i.e., 913 to 3806 mph—1470 to 6125 km/h). Supersonic tunnels work in a manner contrary to the Bernoulli principle as it applies to subsonic and transonic tunnels. As the throat of a wind tunnel constricts, the air velocity is increased for incompressible flow. One would therefore expect the test section to be in this region of maximum velocity. But the air is no longer incompressible as the flow approaches Mach 1. The air compresses and also heats up as it piles up at the constriction. The air actually expands and speeds up following the constriction, and this is where the test object is placed (Fig. 3.3).

Because even the slightest imperfection in the tunnel walls would cause the air to pile up and create shock waves, supersonic tunnels require very smooth interior surfaces.

In spite of the tendency by some to cap the hypersonic range at Mach 15, the hypersonic range only has a low end, which starts Mach 5 or 3806 mph (6125 km/h).Many of the same principles used in supersonic tunnels are also employed in hypersonic tunnels, although they (the hypersonic tunnels) have other problems peculiar to them. For example, a tremendous amount of power (>7.475 kHP/ft^2 [60 MW/m^2] of test section) is required to accelerate the air through the test section. For this reason, hypersonic wind tunnels (as well as supersonic ones) do not operate continuously. Rather, they store up tremendous amounts of compressed air and release it in a brief burst. Another problem is that as the air moves out of the constriction chamber, it cools as its heat energy is converted to kinetic energy. In a hypersonic tunnel, the air can cool so much that it actually turns to liquid. This undesirable occurrence is prevented by deliberately heating the compressed air before releasing it.

3.1.2.3 Classification by Air Pressure

The pressure classification is one of the uncommon classifications of wind tunnels. Tunnels are classified as atmospheric or variable density tunnels. Atmospheric tunnels have open—partially or fully open—test sections. The open test section helps to reduce the build-up of boundary layer and to dissipate the turbulence that may be building up as the flow reaches the test section.

The speed in a wind tunnel can only be increased by so much before the maximum speed is reached or, more seriously, before the tunnel becomes subjected to unacceptable vibration levels. The goal of increasing the speed is to attain the same Reynolds number in the model test as is the case in the object operation condition. This goal can be attained alternatively by increasing the density, since either a higher density or a higher velocity will amount to a higher inertia. (The Reynolds number is the ratio of inertia force to viscous force.) A variable density wind tunnel is essentially a pressure vessel.

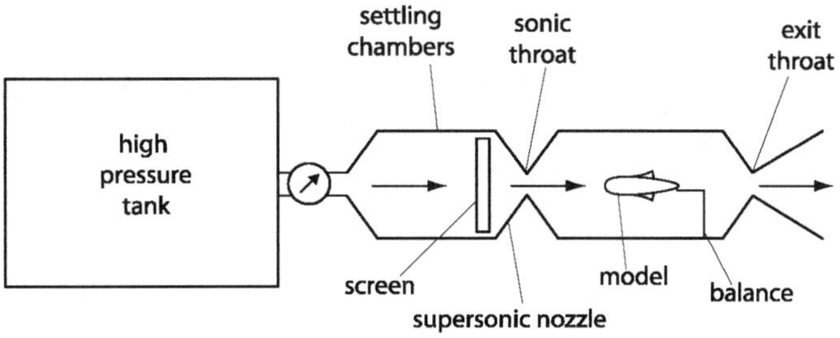

Fig. 3.3 Supersonic wind tunnel

3.1.2.4 Other Types of Wind Tunnel

There are a number of other wind tunnels that fall in a special category of their own. Among them are meteorological tunnel (also referred to as boundary layer tunnel) and shock, plasma-jet, spin, and hot-shot tunnels. Meteorological wind tunnels (boundary layer wind tunnels) are used to study convective boundary layers and other weather phenomena. Shock tunnels are specifically designed for high-speed and high-temperature testing. Gases are compressed and then heated until released to expand rapidly down the expansion chamber. The flow reaches speeds from Mach 3 to well over Mach 20, simulating hypersonic flows, such as is the case in re-entry flights. Like the shock tunnel, a plasma jet wind tunnel is typically used to study the effect of heat exchange on the flow. A spin tunnel is primarily used to study the stall and spin of aircraft and thereby provide useful information for counter-spin flight vehicle design. A hot-shot wind tunnel is one in which electrical energy is discharged into a pressurized arc chamber. The discharge elevates the temperature and the pressure in the arc chamber, causing a diaphragm separating the arc chamber from an evacuated chamber to rupture, and the heated gas from the arc chamber to then accelerate in a conical nozzle, providing flows with very hypersonic Mach numbers for very short (10 to 100 ms) durations.

3.1.3 Wind Tunnel Quality

The quality of a wind tunnel is defined in terms of the range of Reynolds numbers and Mach numbers that it can run. The broader the range, the higher the rated quality of the wind tunnel, with other factors taken into account.

Wind tunnel quality is also expressed in terms of the turbulence level and of the testing equipment. The lower the turbulence level for a given speed, the better is the tunnel rating.

Certainly, the availability and the accuracy of equipment contribute to the appeal of a wind tunnel.

Once we move past the design or functional description of a wind tunnel, the common parameters used to describe a wind tunnel are the maximum speed of the test section, the size of the test section, and the power of the motor that drives the fan. Table 3.1 lists some of the world's wind tunnels famous for automotive testing.

Table 3.1 Automotive Wind Tunnels

Tunnel	Return Circuit Type	Test Section Type	Test Section Dimensions	Maximum Speed, mph (km/h)
RC Canada	Closed	Closed	83.6 m^2	124 (200)
GM USA	Closed	Closed	5.5 × 10.4 m	149 (240)
DNW Netherlands	Closed	Closed	6 × 8 m	137 (220)
Volkswagen Germany	Closed	Open	4.9 × 7.6 m	112 (180)
Lockheed Georgia USA	Closed	Closed	35.1 m^2	252 (406)
MIRA UK	Open	Closed	4.42 × 7.94 m	83 (133)
Daimler Benz Germany	Closed	Open	4.87 × 7.38 m	168 (270)
Fiat Italy	Closed	Open	30 m^2	124 (200)
Ford Cologne Germany	Closed	Open	3.9 × 6.1 m	113 (182)
Nissan Japan	Open	Closed	21 m^2	74 (119)
BMW Germany	Closed	Slotted wall	20 m^2	100 (160)
St. Cyr France	Open	Slotted wall	15 m^2	90 (144)
JARI Japan	Open	Closed	12 m^2	127 (205)

Some of the benefits and disadvantages of the common types of wind tunnels are here itemized in Tables 3.2 and 3.3. They will need to be considered in the selection process of a wind tunnel, and certainly in the early phase of a wind tunnel design.

Table 3.2 Open-Circuit Wind Tunnels

Advantages	Disadvantages
• Inexpensive • Good for flow visualization of smoke and exhaust • Maintains a constant test temperature over the testing duration (if ambient conditions are stable)	• Fluctuating ambient conditions affect free stream air • Noise from the drive fan is not contained • Requires more power (than the closed circuit) to drive the fan

Table 3.3 Closed-Circuit Wind Tunnels

Advantages	Disadvantages
• Requires less energy to drive the fan • Insensitive to ambient conditions	• High cost • Accumulation of smoke (when used as a smoke tunnel) • Build-up of temperature in long duration testing

The boundary layer build-up and temperature rise challenges in the closed-circuit wind tunnel are sometimes addressed with an open jet design for the closed-circuit tunnel. It is particularly crucial in this design to place-hold the model with well-tested, very reliable struts because the opening in the test section makes for a potential hazard that may not be containable within the tunnel. Figure 3.4 shows the open jet wind tunnel.

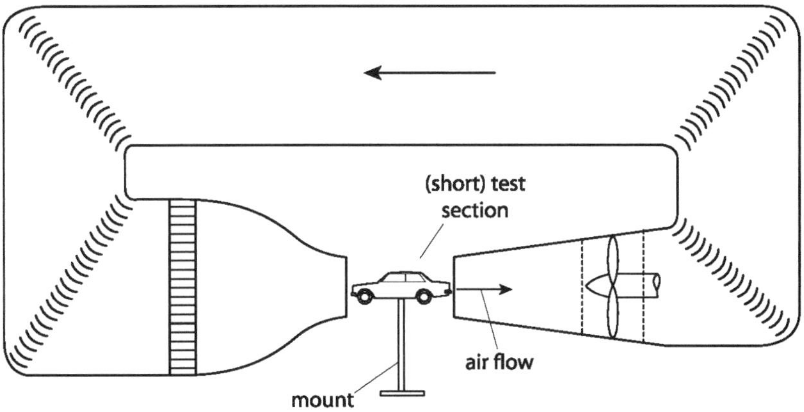

Fig. 3.4 Wind tunnel with open test section

The shape of wind tunnel test sections is usually rectangular, but there is no limitation to such architecture other than the fact that it makes for easy placement of models. In general, the length of the test section is twice the hydraulic diameter ($l = 2b$).

The benefits and disadvantages of common test sections are presented in Table 3.4. One of the types mentioned is the 2-D–3-D combination test section. This design has both the benefit of direct comparison and optimization—the number of tests or configurations that can be performed per unit time. The 2-D section precedes the 3-D, with the latter situated a hydraulic diameter away from the 2-D in this combination design.

Table 3.4 Benefits and Disadvantages of Common Types of Test Section

Test Section Type	Benefit	Disadvantage
Closed jet	Flow boundary well defined—flow quality consistent	High blockage due to build-up of boundary layer
Open jet	Blockage free Easy access to test section from all directions Stable static pressure outside the jet stream	Substantial influence on flow from immediate environment—higher flow instabilities Very short length of test section to minimize influence of environment Requires more power to drive the air
Semi-open jet	Less sensitive to blockage Easy access to test section	Flow instabilities Short length of test section
Adaptive walls	Less sensitive to blockage	Limited access to test section Cost incurred by the additional part
Slotted walls	Insensitive to blockage—can afford a long test section	Poor access to test section Build-up of (static) pressure gradient
Combination (2D-3D)	Tip effect can be quantified in the same test 2-D test or 3-D test can be performed	Longer test section More expensive to build More instrumentation required Higher boundary layer build-up on fixed floor

3.1.4 Wind Tunnel Test Floor Modification

As indicated earlier, the limitation to using a fixed ground plane in a wind tunnel test is the inevitable build-up of boundary layer on the tunnel floor, a phenomenon that does not occur on the road since the head-on air and the floor move as one rigid body on the road. The lower the test piece is to the ground, the greater is the effect of this floor boundary layer on the results of the test. To avoid the influence of the test floor boundary layer, a number of modifications are made to the wind tunnel test section. Such modifications include elevating the ground plane, introducing suction ahead of the model, having a suction plate in the tunnel floor, and introducing a moving belt in the floor.

3.1.4.1 Elevated Ground Plane

The model is mounted on an elevated test floor such that the boundary layer build-up from the wind tunnel floor does not interfere with that due to the car (Fig. 3.5). There is still boundary layer build-up on the elevated floor, but it is very small compared with the build-up on the tunnel floor because the elevated floor is in the inertia-dominant flow regime of the wind tunnel test section.

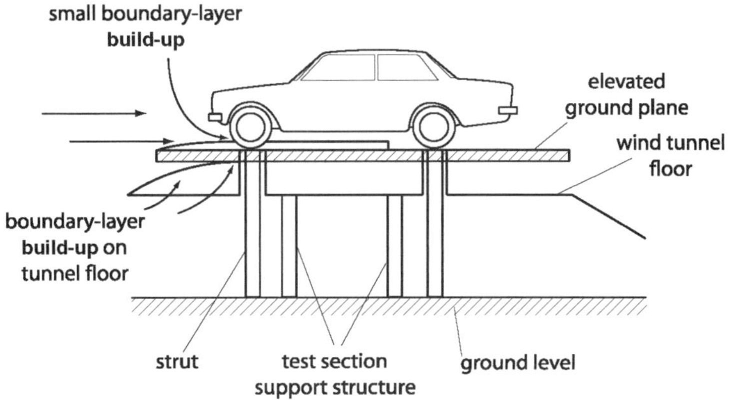

Fig. 3.5 Elevated tunnel floor

3.1.4.2 Suction Ahead of Model

The idea of creating suction ahead of the model (Fig. 3.6) is to dissipate the boundary layer build-up ahead of the model.

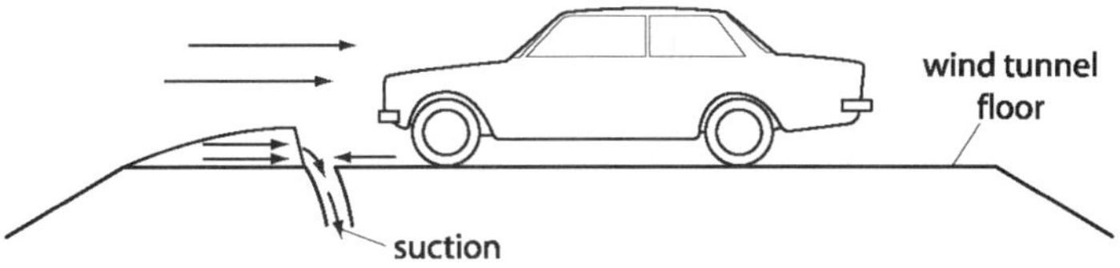

Fig. 3.6 Suction ahead of model

3.1.4.3 Suction Plate in the Tunnel Floor

Having the floor of the test section as a suction plate is more or less an improvement on the design with suction ahead of the model. In this case, the boundary layer build-up on the tunnel floor is repeatedly dissipated so that only the effect of the flow due to the model is recorded (Fig. 3.7).

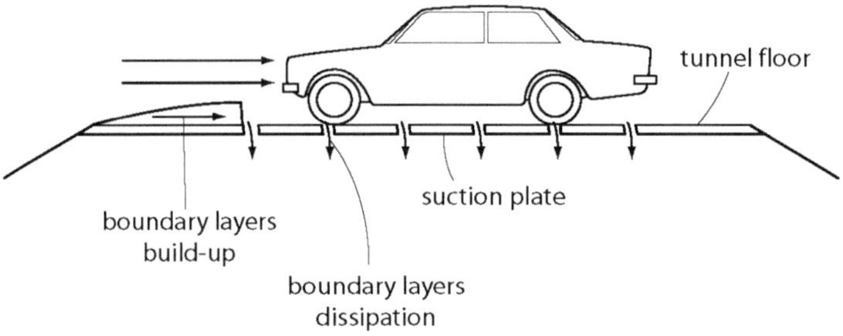

Fig. 3.7 Suction plate in the tunnel floor

3.1.4.4 Moving Belt

The idea of a moving belt test section is to disallow the build-up of boundary layer over time at some fixed distance from the test section entrance [3-7]. A starting boundary layer on the test floor is carried away by the moving belt, and another boundary layer starts as the belt turns. This way, the moving floor simulates a true-ground relative to the model (Fig. 3.8).

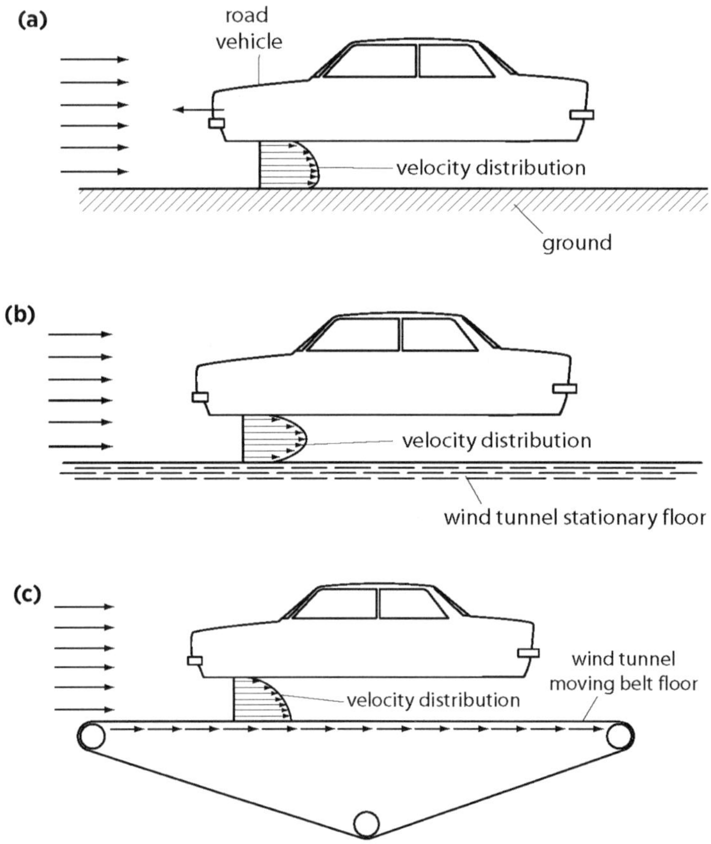

Fig. 3.8 Effect of a moving-belt test floor

A combination of these designs is used in some tunnels, depending on the degree of accuracy sought.

3.1.5 Wind Tunnel Model Size and Wall Correction

In Section 1.7, Fig. 1.16 describes the flow over an airfoil in an unconstrained environment. Even in that case, the streamlines around the object are constrained to be different from the free stream. The so-called unconstrained environment is actually constrained, but the constraint is at infinity, practically. At some distance to the object, the effect of the constraint will become manifest. From that region on, the closer the bounds are to the object, the greater will be the wall effect. This is the situation with an object in a wind tunnel. The size of the model that can be used in a wind tunnel is limited by the size of the wind tunnel test section. Since the ideal model is a full-scale model, the closer the model is to a full scale the more desirable it is. If the model is not large enough, it may not include critical details. However, the larger test section-to-model size ratio, the more reliable are the test results. If the test section is not sufficiently wide, the flow will move faster and will register a higher lift (and drag) reading. But increasing the size of the test section incurs a price: the larger the test section cross-sectional area, the larger is the power requirement (higher velocity), and the greater the cost. The optimum is a sufficiently wide test section and a model large enough to capture the real-life details intended to be captured, yet small enough to be free from wall influence caused by the test section.

The term *blockage* describes the interference in the wind tunnel test section due to the obstruction of flow. The obstruction comes in the forms of solid blockage and wake blockage. Solid blockage is due to the test model whereas wake blockage is due to the solid wall of the test section. Solid blockage is defined as the ratio of the model frontal area to the test section cross-sectional area. It is recommended that the solid blockage for vehicle testing be no more than 10%. To give a quantitative illustration for the effect of blockage, consider Fig. 3.9(a).

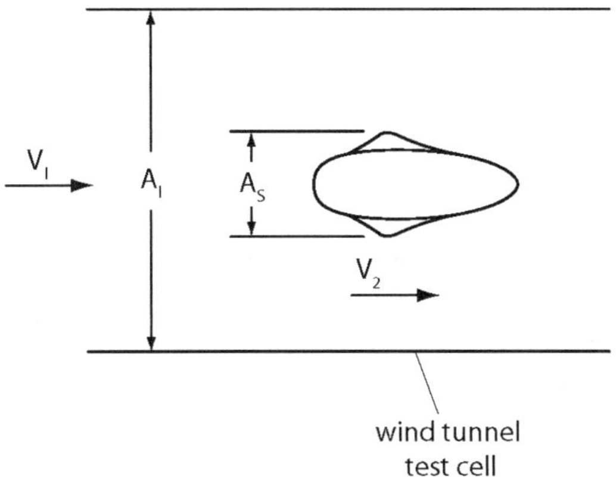

Fig. 3.9(a) Solid blockage

Define $A_2 = A_1 - A_s$.

From continuity of incompressible fluid

$$v_1 A_1 = v_2 A_2$$

where

v_1 is the recorded speed, as read in the test section; and
v_2 is the true speed 'experienced' by the model the model, but impractical to measure.

Let $D_{measured}$ be the apparent or measured drag, and D_{real} the real drag. D_{real} is the drag that would be recorded if the model was on the road, with no bounding walls, and it would be

$$D_{real} = \frac{1}{2}\rho v_1^2 C_{D_{real}} A_s \quad (3.1)$$

$D_{measured}$ is, however, the drag recorded in the wind tunnel, and its true value is

$$D_{measured} = \frac{1}{2}\rho v_2^2 C_{D_{real}} A_s \quad (3.2)$$

Because the drag increases with speed, $D_{measured}$ (Eq. 3.2) is greater than D_{real} (Eq. 3.1).

From Eq. 3.2, true $C_{D_{real}}$, in terms of the measured drag, is

$$C_{D_{real}} = \frac{D_{measured}}{\frac{1}{2}\rho v_2^2 A_s} \quad (3.3a)$$

which, from Eq. 3.1, is the same as

$$C_{D_{real}} = \frac{D_{real}}{\frac{1}{2}\rho v_1^2 A_s} \quad (3.3b)$$

The computed drag coefficient is defined as

$$C_{D_{computed}} = \frac{D_{measured}}{\frac{1}{2}\rho v_1^2 A_s} \quad (3.4)$$

Dividing Eq. 3.3a by Eq. 3.4

$$\frac{C_{D_{real}}}{C_{D_{computed}}} = \frac{v_1^2}{v_2^2} \quad (3.5)$$

Applying continuity, and rearranging Eq. 3.5, we obtain

$$C_{D_{real}} = C_{D_{computed}}\left(1-\frac{A_s}{A_1}\right)^2$$

which, when expanded, yields

$$C_{D_{real}} = C_{D_{computed}}\left(1-2\frac{A_s}{A_1}+\left(\frac{A_s}{A_1}\right)^2\right) \tag{3.6}$$

If we impose the industry standard, that is $A_s \le 0.1A_1$, in (3.6), we would have the real drag coefficient, $C_{D_{real}}$ to be as low as 81% of the computed value. It gets even lower when the effect of wake blockage is included (Fig. 3.9(b)).

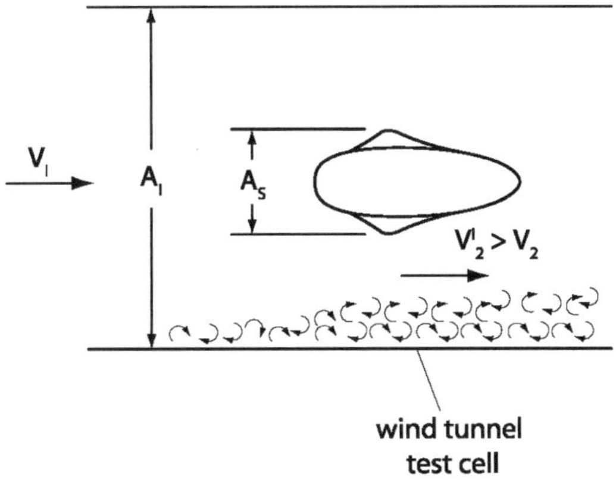

Fig. 3.9(b) Blockage (solid and wake)

Because solid blockage results in an increased dynamic pressure, it affects all the aerodynamic forces. Wake blockage, however, primarily influences drag. A generally accepted first pass expression for the wall correction of the aerodynamic coefficients is of the form

$$C_{D_{real}} = \frac{C_{D_{computed}}}{\left(1+\frac{1}{4}(A_s/A_1)\right)^2} \tag{3.7}$$

The larger the blockage, the further away from reality is value of coefficient obtained from the measured aerodynamic forces.

One might think that the problem of blockage is completely removed in an open test cell design, but that is not the case. Because the flow in the tunnel is at a substantially higher speed than ambient, there is a significant pressure difference between the open test cell and the immediate environment; hence, a virtual pressure wall that seeks to push into the test cell. In accordance with Bernoulli's flow principle, the higher the tunnel speed, the greater is the pressure difference, and the more the virtual wall wants to push into the test section. Nonetheless, the blockage in an open test cell is very little in comparison with that in a closed test cell.

3.1.6 Mounting Devices

In selecting the mount for a model in the test section, there are three primary goals: that the mount adequately holds the model in place, that the mount does not add to the drag measured for the model, and that it is easy to install and to disassemble. There are several ways to mount the model in the test section, depending on the wind tunnel design. In a proposed wind tunnel design, it is worth taking into account the types of mount that will be used, as this will influence the particular wind tunnel design.

3.1.6.1 Model Placement

To understand the importance of positioning a model in the wind tunnel, we will address the flow cross section in a wind tunnel. As shown in Fig. 3.10(a), air flow is not uniform across a section of the test cell, especially in a closed cell. The highest speed is in the central region of the test section, and it decreases toward the wall of the test cell; this disparity in velocity increases with the Reynolds number. The same is true of an open jet, but the drop in velocity is much more pronounced in a closed jet test section. It is therefore important to put the model in the center of the test section, not only because that is where the speed is highest—a desirable feature—but also because the test results are repeatable and the test is least influenced by the boundary layer effect of the tunnel wall.

Another factor to consider in placing a model is the interference effect of protrusions in the model. The correction factor, $\beta = \sqrt{(1 - M^2)}$, to address the interference effect is based on the assumption that only the longitudinal flow is subject to interference. That assumption is functionally true only if the model is mounted in the lateral center of the wind tunnel [3-16].

It is common to chamfer or round the corners of test sections to avoid the tremendous drop in velocity that manifests in the corners. In Fig. 3.10(a), v_c can be as low as 0.7 v_∞. Fig. 3.10(b) provides a more uniform velocity distribution. Though it does not eliminate the wall effect, it averts the corner effect and its consequential turbulence build-up.

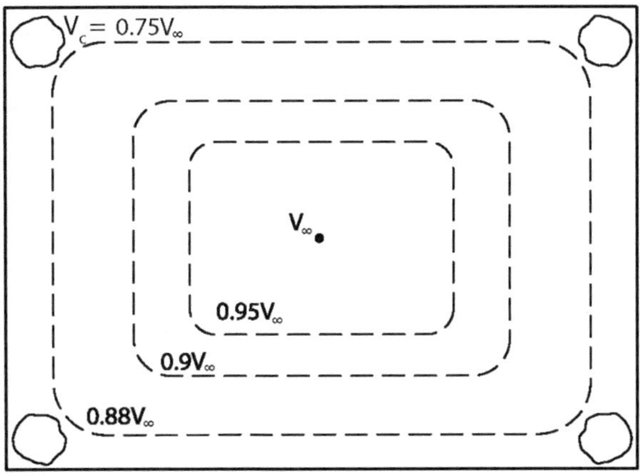

Fig. 3.10(a) Cross-section of velocity distribution in the wind tunnel test cell

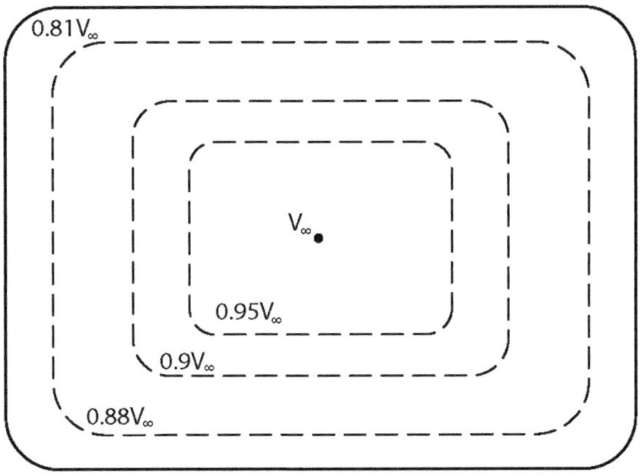

Fig. 3.10(b) Improved test cell

3.1.6.2 Internal Balance

The measuring devices in an internal balance are embedded in the model (Fig. 3.11). The attached instrumentation sends information from the model to a data recording device. Models in which internal balances are used are necessarily big compared with those that would be used with external balances. This is to allow for the tubes, the pressure probes, the strain gages, and wires that need be embedded in the model and to provide a path within the model for connecting wires to the data acquisition port, which is outside the test cell. Although generally more expensive than the external balance, it is compact and more accurate.

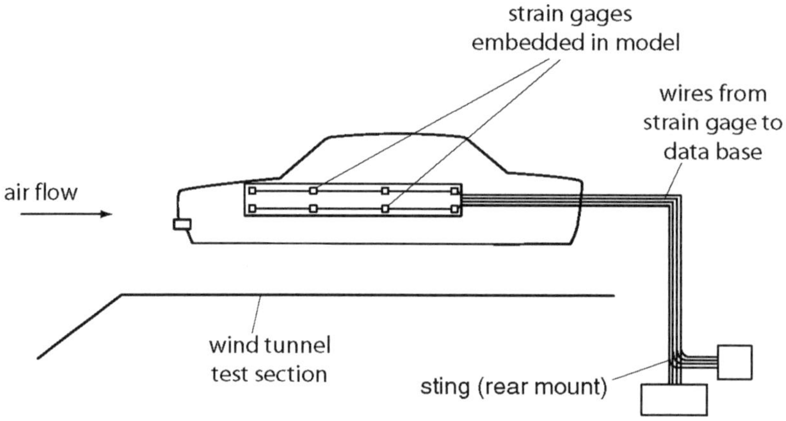

Fig. 3.11 Internal balance

3.1.6.3 External Balance

When external balances are used, the support is typically outside the model, and the measuring devices are outside of the test cell. Data is obtained indirectly by the transmission of aerodynamic load information from the model to the external balance.

External mounts are of various designs, depending on the goal of the test and the architectural constraints in and around the wind tunnel. Some forms of external mounts are direct strut mounts, rotating tables, rear-mount stings, and ceiling stings (Fig. 3.12 (a–e)). The strut mounts can be three-strut mounts, two-strut mounts, or single-strut mounts. The three-strut mount is the most stable, but it is expensive and cumbersome. It also generates substantial flow interference, with the leading struts causing primary interference, and the trailing one adding a secondary interference to its primary contribution. With the two-strut mount, there is less interference, but it is not as rigid as the three-strut so it may not be as reliable in performance tests. The single-strut mount is the least intrusive to air flow and the least expensive but less rigid than three-strut or two-strut mounts. Its top usually requires some modification to a table top to keep the model stable. It is more fitting for internal balances, especially in the form of a rear-mount sting or ceiling sting. The sting mount is a form of single-strut mount, but it has less interference with the model flow field than the standard single-strut mount. The ceiling sting has an additional benefit of negligible contribution to wake and to trailing vortices, but the model size used in ceiling sting mounts is usually smaller than those used in other mounts.

Rotating table mounts are particularly valuable in cross-wind experimentation. The intent of rear-mount stings is to eliminate the contribution of the strut to the measured model loads.

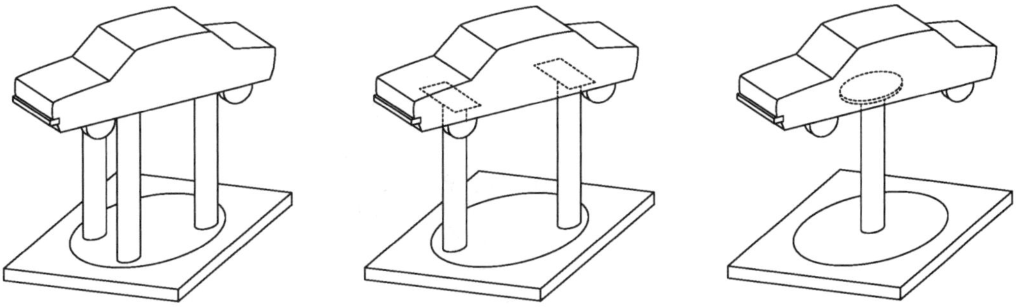

(a) 3-strut balance **(b)** 2-strut balance **(c)** single-strut balance

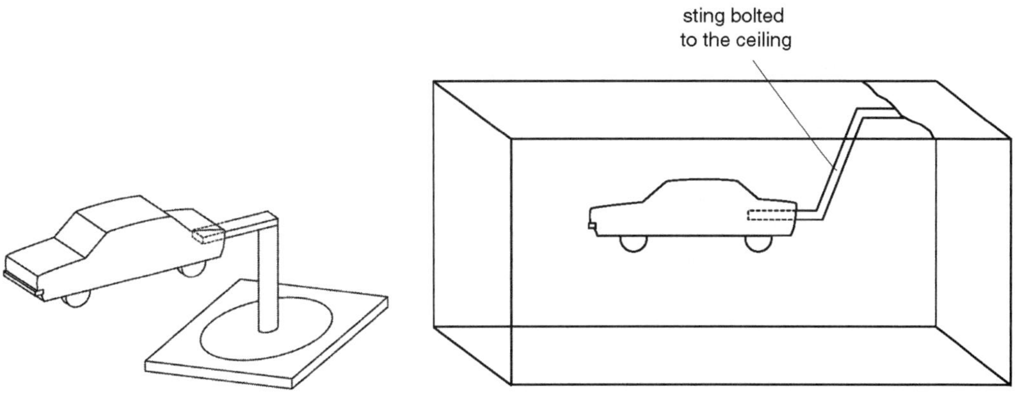

(d) rear-mount sting **(e)** ceiling sting

Fig. 3.12 External balance

3.1.7 Measured and Computed Parameters

The force and moment data are the most commonly collected direct data from wind tunnel tests. These values are then used to compute their nondimensional counterparts, the coefficients of the forces, and the moments, as shown in Section 1.8.

$$\text{Lift coefficient, } C_L = \frac{\mathrm{L}}{\frac{1}{2}\rho \mathrm{v}^2 A}$$

$$\text{Drag coefficient, } C_D = \frac{D}{\frac{1}{2}\rho \mathrm{v}^2 A}$$

$$\text{Yaw coefficient, } C_Y = \frac{Y}{\frac{1}{2}\rho \mathrm{v}^2 A}$$

where Y is the yaw or the side force.

Moments:

$$\text{Pitching moment, } C_M = \frac{M}{\frac{1}{2}\rho v^2 Ac}$$

$$\text{Rolling moment, } C_R = \frac{R}{\frac{1}{2}\rho v^2 Ac}$$

$$\text{Yawing moment, } C_N = \frac{N}{\frac{1}{2}\rho v^2 Ac}$$

where, as shown in Fig. 1.19

M is the pitching moment, about the lateral axis;

R is the rolling moment, about the longitudinal axis; and

N is the yawing moment, about the vertical axis.

The pressure coefficient, C_p, is defined as the ratio of the pressure difference (between a port and the ambient) to the free stream dynamic pressure. This will be addressed more in Section 3.1.8.

$$C_p = \frac{p - p_\infty}{\frac{1}{2}\rho v_\infty^2} \tag{3.8}$$

Lift-to-drag (L/D) ratio. The lift-to-drag (L/D) ratio is a measure of vehicle performance. For an aircraft, the higher the L/D ratio, the better is the vehicle performance. For ground vehicles, the higher the negative L/D ratio, the better is the vehicle stability since a high downforce (negative lift) is the goal for vehicle stability.

3.1.8 Instrumentation

In addition to the load and moment measuring gages (Fig. 3.11), the surface pressure distribution is measured to determine the pressure force—a major contributor to the aerodynamic load—on the vehicle.

It is, in fact, more convenient to measure the pressure than it is to measure the velocity. As in the case of the lift coefficient, the nondimensional pressure coefficient is a more convenient parameter to use than the actual pressure. Knowing the pressure coefficient, the velocity can be computed using Bernoulli relationship [3-17, 3-18].

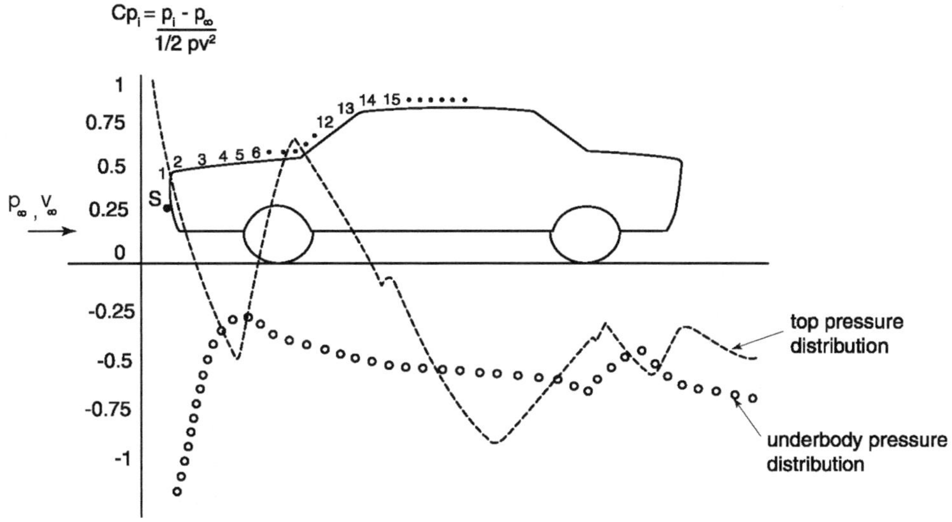

Fig. 3.13 Typical pressure distribution along a car's longitudinal axis

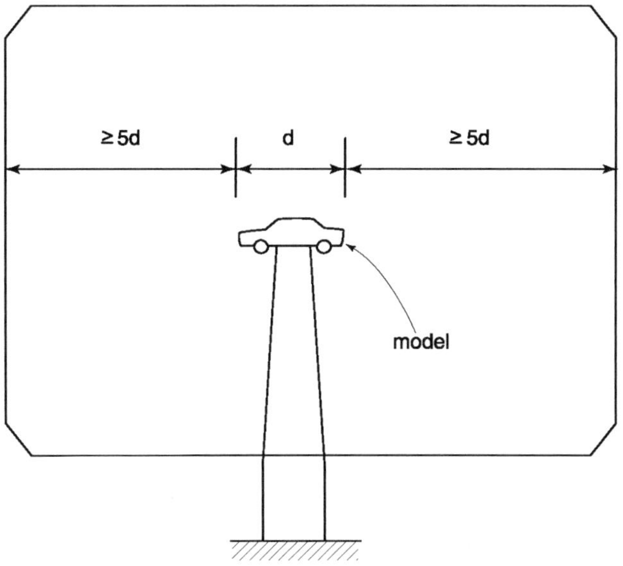

Fig.3.14 Positioning a model in the wind tunnel

Recall (1.15b)

$$p_1 + \frac{1}{2}\rho v_1^2 = p_2 + \frac{1}{2}\rho v_2^2$$

If we apply this relationship between free stream condition and stagnation point S in Fig. 3.13, we obtain

$$p_{total} = p_\infty + \frac{1}{2}\rho v_\infty^2 \tag{3.9}$$

where

p_{total} is stagnation pressure.

The dynamic pressure therefore becomes the difference of the total and free stream values, from Eq. 3.9. Once the static pressure at a point, *i*, on the surface of the vehicle is measured, the pressure coefficient is obtained since the free stream parameters are already known.

Equation 3.9 is the common form of expression for the pressure coefficient. The same can be obtained using velocity relationship. Again, if we applying Bernoulli's relationship to the free stream and the point, *i*, we obtain

$$p_i + \frac{1}{2}\rho v_i^2 = p_\infty + \frac{1}{2}\rho v_\infty^2$$

which yields

$$p_i - p_\infty = \frac{1}{2}\rho(v_\infty^2 - v_i^2) \tag{3.10}$$

Replacing $p_i - p_\infty$ in Eq. 3.8 with the right-hand equivalent in (3.10), we have

$$C_p = \frac{\frac{1}{2}\rho(v_\infty^2 - v_i^2)}{\frac{1}{2}\rho v_\infty^2}$$

which, when reduced, gives

$$C_p = 1 - \frac{v_i^2}{v_\infty^2} \tag{3.11}$$

Depending on the data available—local static pressure or local velocity—either Eq. 3.8 or Eq. 3.11 can be used to calculate the pressure coefficient at any point on the vehicle surface.

There are various means of collecting pressure data. They include pressure port, anemometer, wake rake, pitot probes, and transducers. Pressure ports are holes drilled flush to the surface of the model. Pressure ports provide the local static pressures that when integrated give the forces on the ported surfaces and implicitly gives a picture of the distribution of forces on the model. Though hot wire anemometers are typically used to measure the speed in a wind tunnel, pressure transducers measure the pressure. A wake rake is a qualitative device. It is used to determine the wake characteristics of a model. Just as pressure ports provide pressure data, pitot probes provide velocity data. They are used to measure the velocity gradients, which are then used to determine the drag on a model.

3.1.9 Flow Visualization

Smoke tunnels are used mainly for flow visualization. The smoke traces indicate the direction and smoothness of air flow on the vehicle body. Because they provide a quick, qualitative indication of the flow pattern around an object, the stylist is able to make improvements on-the-fly. Although computerized data acquisition systems are sometimes incorporated [3-18], flow visualization is primarily qualitative.

3.1.10 Scale Model Testing

It is important to be able to determine the type of model required to address a particular aerodynamic issue or a set of issues and, using dimensional similarity, to size the model appropriately for a given wind tunnel. Ideally, the details of a model should reflect the same details in the real vehicle. However, this can quickly become a costly venture because of the many details in a vehicle, and because the smaller the scale (and scale models are often small—about 1/30th full scale) the more taxing it is to replicate such details as contours and protrusions. Therefore, some details may be let go in the scale model if it has been estimated that such exclusion does not significantly affect the test result or the application of it in the real vehicle.

3.1.11 Scale Effect

Even when we have a geometrically similar scale model, there are some physical phenomena that cannot be scaled. For example, the transition from laminar flow to turbulent flow cannot be scaled. As a result, the flow over a reduced scale model in the wind tunnel transitions to turbulence much later than the transition point in the real vehicle. In fact, for very small scale models, it is possible that there is no transition to turbulent flow. This is usually corrected by tripping the flow on the scale model. The geometrically equivalent transition point in the model is roughened so that the flow can start to be turbulent at that point.

3.1.12 Effect of Wheels

The contribution of wheels to vehicle drag will be treated to a greater extent in chapter 5. In this chapter, we merely seek to address the modeling of wheels in wind tunnel testing, and its effect on the wind tunnel result. The turning of the wheels affects the motion of a vehicle, yet wind tunnel tests are often performed with the wheels fixed to the body of the vehicle. Although this is done for model simplification, it takes away from the accuracy of the test result. Here, we will consider three wheel formats: fixed wheels on test floor, floating fixed wheels, and rotating wheels.

When the wheels are stationary on the test floor (Fig. 3.15), they obstruct air flow (the front wheels being more obstructive than the rear ones) and add to the vehicle drag. The flow also causes an increase in lift. The result in operation is a reduction in downforce. (Keep in mind that an increase in ground force is desirable for vehicle stability.)

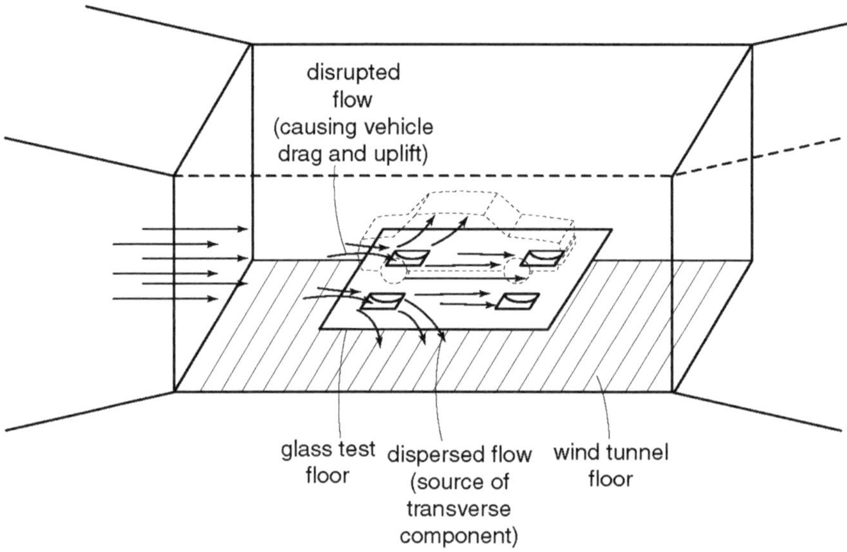

Fig. 3.15 Stationary wheels

To better represent the true occurrence without a rotating set of wheels, the fixed wheels are raised from the test floor so that there is no pronounced longitudinal blockage of the flow (Fig. 3.16(a)). The air flow passes between the wheels and the test floor, essentially maintaining its flow pattern, and avoiding pressure build-up at the leading edge of the wheel. For a given gap, Δh, between the wheels and the test floor, both the drag and the lift (upforce) reduce with speed up until a critical point after which there is no further benefit in increasing the speed. The same trend in benefits is the case when Δh is increased up until δ, the latter being the boundary layer thickness at a given tunnel speed. It is impractical to estimate δ each time a test is run, so a standard value, δ_m, is obtained once and thereafter used for all tests in a given tunnel. δ_m is the boundary layer thickness due to the test floor when the wind tunnel is run at its maximum speed. An experimental design to compare of Figs. 3.16(a) and (b) will show that the reduction in drag and in upforce in Fig. 3.16(a) is due exclusively to the gap between the wheel and the floor. A further comparison of Figs. 3.16 (b) and (c) will show that most of the reduction takes place in the front wheels, for not only are the lift and drag coefficients reduced in Fig. 3.16(c), they are close to the values obtained in Fig. 3.16(a).

Stationary wheels produce a larger wake region near the ground than do rotating wheels at the same close proximity to the test floor. Beyond δ, there is no noticeable difference in the wake size between the wakes generated by stationary and rotating wheels.

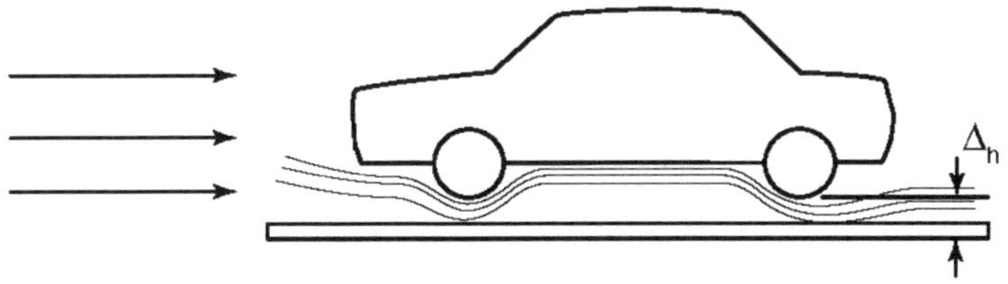

Fig. 3.16(a) Floating fixed wheels

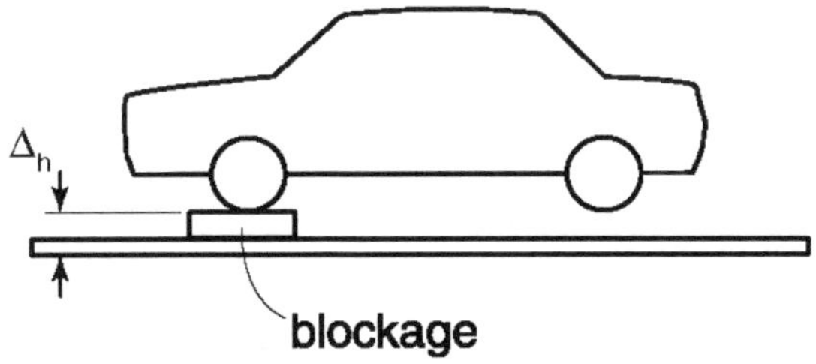

Fig. 3.16(b) Blockage at the front wheels

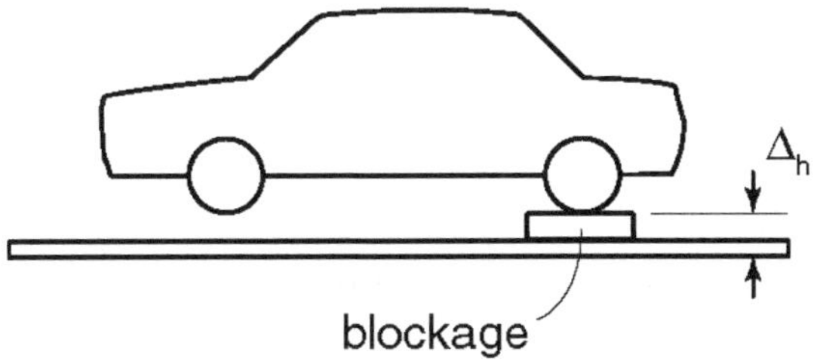

Fig. 3.16(c) Blockage at the rear wheels

Rotating wheels better represent the real-life picture than fixed wheels, although it is more expensive to incorporate rotating wheels in the scale model. Because the wheels rotate sympathetically with the direction of air flow (Fig. 3.17), the air speed at the wheels is higher than is the case for fixed wheels, and the pressure is lower. As a result, there is a higher downforce. As with fixed wheels, the downforce increases with speed and with the height of the wheel from the test floor. Also, the drag decreases, albeit minimally, with the height of the wheel from the test floor but increases with speed.

When compared with the case of a wheel on a stationary test floor plane, the upper separation point of the moving plane shifts forward. This forward migration of the separation point would not be observed for a wheel in a wheel-well. Also, the pressure coefficient, C_p, at the wheel-ground contact initiation point is higher for the case of the moving ground, rising to about twice the value for the stationary floor [3-19].

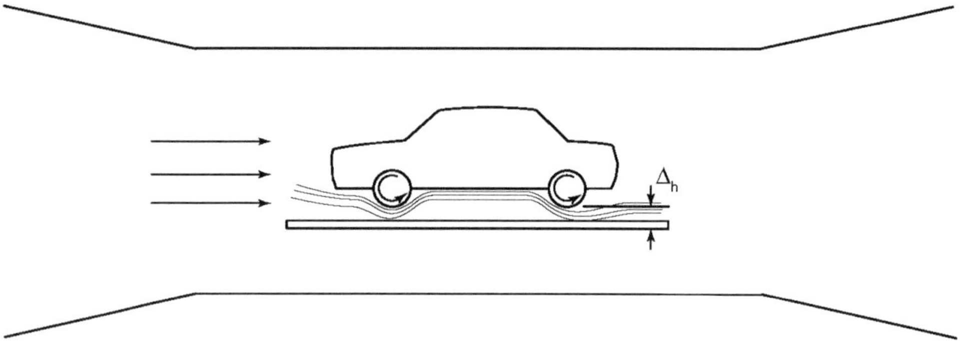

Fig. 3.17 Rotating wheels

Adding wheel-wells to the basic auto body causes a notable change in the drag and lift of the vehicle. Both lift and drag are increased by the addition of wheels and wheel-wells to the basic vehicle frame. The wheel produces a flow that interferes with the vehicle body flow. This causes an increase in the drag. The lift is due directly to the effect of the flow from the wheel on the wheel-well.

The closest ground simulation seems to be a moving ground plane and rotating wheels to give a close-to-real ground and wheel effect. But it has already been established that a moving ground surface does not significantly improve wheel simulation and that the poor correlation increases, to an extent, with the gap between the wheel and the ground plane [3-19, 3-20, 3-21, 3-22]. A fairly accurate simulation can be obtained with a stationary wheel and ground plane, keeping a small gap between the wheel and the ground plane [3-23].

3.1.13 Similarity

The concept of similarity is the basis of model testing, whether wind tunnel testing, water tunnel, or any other form of testing that replaces the real object with a scaled model of the same. Geometrical similarity calls for a physical model that is similar to but (usually) smaller than the real object. Thanks to the principle of dynamic similarity, we are able not only to construct meaningful model tests but to interpret and correctly apply the result to the full-scale object. In dynamic similarity, flow conditions are replicated in the model so that the numerical values of the nondimensional parameters used are the same for the model as for the real vehicle. The most common of the similarity parameters is the Reynolds number.

3.1.14 Reynolds Number

The Reynolds number is the ratio of inertial force to viscous force (Re = $\rho vd/\mu$). Consider a flow at a very low Reynolds number. This implies that the viscous force significantly dominates the inertial force and, as a result, the inertial force is negligible. The force balance (pressure force + viscous force = inertial force) therefore shows that, for all practical purposes, the pressure force pushing the fluid balances the viscous force, which resists the fluid deformation. On the other hand, when the Reynolds number is very high, it is the viscous force that is negligible in comparison with the inertial force, and, in this case, a balance of forces exists between the pressure force and the inertial force, the force resisting acceleration.

It is important to note that the Reynolds number is not an indication of the viscosity of the fluid or of its inertia. Rather, it is an indication of the influence or the strength of the inertia relative to the fluid viscosity. A fast-moving, highly viscous fluid with a high Reynolds number may have a negligible viscous effect in comparison with a sluggish stream.

As the Re increases, the boundary layer grows less rapidly. However, the viscous boundary layer never disappears because the velocity at the wall remains at zero (no slip).

Other similarity parameters include Mach number, Prandtl number, Strouhal number, Froude number, and Eckert number. Their applications are listed in Table 3.5.

Table 3.5 Similarity Parameters

Similarity Parameter	Symbol and Definition	Field of Use	Comment
Reynolds number	Re = $\rho vd/\mu$	Dynamic similarity	Smaller model requires higher speed in the same proportion, e.g., one-third scale model requires 3 times air speed. Air flow must be below shock condition.
Mach number	$M = u/a$	Compressible flow	High speed, variable density
Prandtl number	Pr = $C_p\mu/k$	Convective heat transfer, thermal boundary layer	Ratio of momentum diffusivity (kinematic viscosity) and thermal diffusivity
Strouhal number	$St = \omega L/u$	Unsteady flows, oscillating flows, wakes, vortices	Ratio of inertial forces due to the unsteadiness of the flow or local acceleration to the inertial forces due to changes in velocity from one point to another in the flow field
Eckert number	$Ec = \dfrac{u^2}{C_p(T_w - T)}$	High-speed flows with heat exchange	Ratio of kinetic energy of flow to enthalpy
Froude number	$u/\sqrt{gL}$	Wave and surface flows	Ratio of inertial forces to gravitational forces. Typically used in water tunnel experiments.

where

a = speed of sound
g = acceleration due to gravity
u = flow velocity
L = length scale
k = coefficient of thermal expansion
ω = oscillation frequency
T_w = wall temperature.

3.1.15 Estimating Full-Scale Values from Model Data

Due to power limitations or excessive structural vibration, the Reynolds number at which a test is run is often lower than the Reynolds number of the vehicle's real-life operation. Fortunately, the results of wind tunnel tests are comparable to real-life events, even when the Reynolds number of the wind tunnel run is lower than that of the true event. The practice is therefore to run the tunnel at the highest possible speed, taking structural vibrations into account.

As addressed in Section 3.1.5, blockage in the wind tunnel should be determined, preferable before a model is built, so as to ensure that it is below 10%.

Consider the following example:

A one-fifth scale model of a vehicle is to be tested in a square test cell wind tunnel of 1.2 m side. The projected frontal area of the vehicle is 3.5 m^2. The blockage for this model-tunnel combination is

$$\frac{\text{Cross sectional area of vehicle}}{\text{Cross sectional area of wind tunnel}}(\text{scale})^2 \times 100\%$$

which gives

$$\frac{3.5}{(1.2)^2}\left(\frac{1}{5}\right)^2 \times 100\% = 9.72\%, \text{ just below the 10\% threshold.}$$

The acceptance of this value presumes that the model is centralized (Fig. 3.10) and that the lateral distance from the models boundary to the tunnel wall is five times the width of the model (Fig. 3.14).

3.2 Track and Road Testing

In road testing, we have the benefit of testing the real thing. Unfortunately, even this does not provide the utopia we seek in vehicle testing. There are four main disadvantages. The first concerns instrumentation. Most of the instruments needed are cumbersome. Portability was not an included parameter in their original design. Also, room must be made in the vehicle for the instrumentation. Sometimes, this required instrument space competes with crew accommodation in the vehicle. In addition, more than one set may be required to adequately compare two design styles of a prototype. The second disadvantage is the re-instrumentation required for changes made to a vehicle. Even slight changes may require re-instrumentation, especially if the change is made at a data point. A third disadvantage is the inability to control the environment. Though tracks can be designed to simulate road conditions of interest, it is difficult (if not impossible) to control the weather. This makes it impossible to predict the wind direction or its speed to which the road vehicle will be subjected, hence the challenge of repeatability. The fourth, which is perhaps the primary challenge, is that a real-life vehicle must have already been built. The cost and labor of this, over building a reduced-scale wind tunnel model or a true-dimension computational model, makes road testing unpalatable for concept vehicles. It is usually used for improvement on existing designs [3-24, 3-25, 3-26, 3-27, 3-28].

In road testing, because of the quick response required to the ever-changing road conditions, it is paramount that the tester be familiar enough with instruments used in road testing to be able to make quick adjustments to prevailing road conditions. The operator must understand methods of data acquisition and be able to interpret track data for decision making in aerodynamic optimization of vehicles.

3.2.1 Aerodynamic Data Acquisition

Both measuring instruments and measuring techniques are as important as the parameter to measure for any data acquisition. Load distribution, aerodynamic pressure, and air temperature are crucial independent parameters in computing a vehicle's aerodynamic characteristics. A measure of the load distribution is obtained from knowing the suspension travel, and measurements of pressure and temperature are obtained from probes mounted at strategic locations on the vehicle.

3.2.1.1 Suspension Travel

The suspension travel of each wheel is measured, and with one of the wheels as reference, the relative travel of each of the other wheels is determined. This gives a measure of the relative load distribution, hence the downforce, in the vehicle.

3.2.1.2 Mounting Probe

A mounting probe is a probe on the vehicle that measures the dynamic head. Probes are also installed to measure the air temperature. The pitot static tube is usually used for pressure measurement. For corroboration, the same data is sometimes measured by two different devices.

3.2.1.3 Data Acquisition

Tire rolling resistance and surface pressure are among the data that can be determined directly in road tests. Surface pressure is obtained by taking measurements from the pressure port drilled flush with the surface of the vehicle. One of the methods to measure the rolling resistance is shown in Fig. 3.18. The test vehicle is shrouded in a trailer and sealed as much as is feasible so that no extraneous air influences the recorded axial load. An axial load cell attached to the vehicle records the rolling resistance with varying vehicle speed. This (rolling resistance) is then subtracted from the total drag. The result is the vehicle aerodynamic drag. A common method of measuring the total drag is the coast-down method. In coast down, the vehicle is accelerated to a high speed—usually over 100 mph (160 km/h). That top speed is referenced as the initial time, $t = 0$. The vehicle is then slowly decelerated while the speed is recorded at small time intervals. From Fig. 3.19, the approximate deceleration at a vehicle speed is obtained, and, when multiplied by the vehicle gross mass (including the mass of a full tank of fuel), the vehicle total drag, $D = m_{total}(-a)$, is calculated.

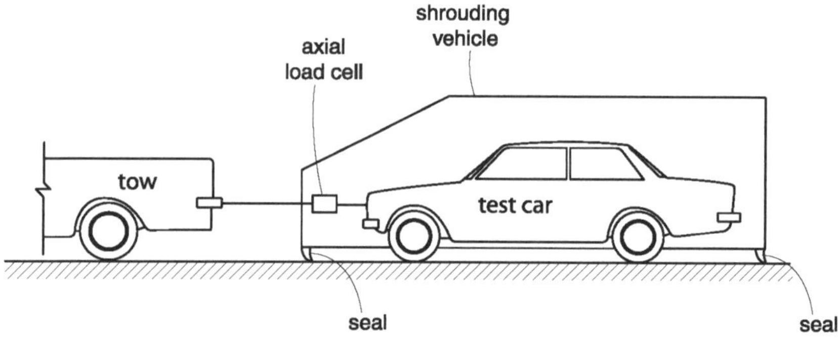

Fig. 3.18 Measurement of vehicle rolling resistance

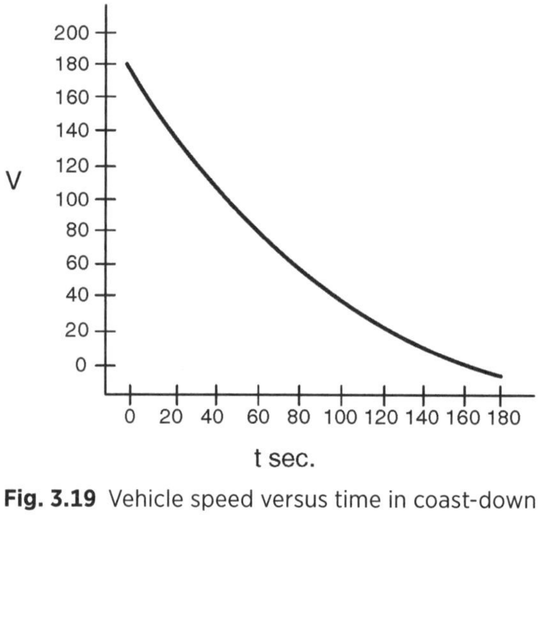

Fig. 3.19 Vehicle speed versus time in coast-down

3.2.2 Data Standardization

For comparison, the results obtained are converted to standard temperature and pressure (STP) values by correcting for temperature and pressure. The coefficients of lift and drag are obtained by dividing the lift and drag, respectively, by the vehicle projected frontal area. The pressure coefficient is obtained by dividing the measured surface pressure by the free stream dynamic pressure, as shown in (3.8) and (3.11).

Summary

Experimental tests are valuable to establish a theory, to verify a theory, and to validate a computational result. There are two main types of experimental tests for ground vehicles: wind tunnel tests and road tests. Wind tunnel testing is usually done on scaled-down models and road tests are performed on full-size vehicles. They both have advantages and shortcomings.

For wind tunnel test models, there is the disadvantage of a scaled-down model failing to fully replicate the geometric and dynamic features of the full-scale vehicle. Also, the wind tunnel speed is generally limited, resulting in a lower test Reynolds number than in the full-scale vehicle test. Road tests have the disadvantage of being at the mercy of the weather and the road conditions. Although the full-scale vehicle (or component) is being tested, it is often difficult to replicate the test result because of environmental variability.

Key Terms

Blockage: Interference in the wind tunnel test section due to the obstruction of flow. There are two types of blockage, solid blockage and wake blockage. Solid blockage is due to the test model and wake blockage is due to the solid wall of the test section. Solid blockage is defined as the ratio of the model frontal area to the test section cross-sectional area. It is recommended that the solid blockage for vehicle testing be no more than 10%.

L/D ratio: Lift-to-drag ratio is an indication of the dynamic stability of a vehicle. For ground vehicles, the numerically higher the negative L/D ratio, the more stable is the vehicle.

Similarity: The principle of creating a nondimensional equivalence between an object and its scale model.

Transonic wind tunnel: The range in transonic wind tunnels is Mach 0.8 to 1.2 (609–913 mph, 980–1470km/h). Slotted walls are usually included in the design of transonic wind tunnels to reduce shock reflection from the walls of the test section. Mach number starts to be important for transonic flows due to the development of shock waves.

Wind tunnel: Device primarily used for aerodynamic testing of a component or vehicle

Wind tunnel quality: An expression of the range of Reynolds numbers and Mach numbers that a wind tunnel can run. In general, the broader the range, the higher is the rated quality of the wind tunnel.

Review Questions

1. Label and describe the sections of a closed circuit wind tunnel.
2. Name and describe four types of wind tunnel based on architectural classification.
3. Name and describe four types of wind tunnel based on classification by speed.
4. What is meant by wind tunnel quality?
5. Compare the a moving belt test section with a suction plate test section.
6. Define (a) solid blockage, (b) wake blockage.
7. Name and describe (a) the measured parameters and (b) the computed parameters in wind tunnel testing.

Application Questions

1. Given a choice between an open jet test section and a closed jet test section, how would you determine which to choose?
2. How would you reduce the effect of dynamic blockage in model testing without reducing the air speed in the tunnel?

Exercise

In Fig. 3.9(a), if A1 is 1 m^2, and the model cross section is 625 cm^2, what is the speed experienced by the model when the tunnel is running at 500 km/hr? By how much is the measured drag greater than the real drag (assuming sea level driving condition)? What is the ratio of the real drag coefficient to the computed drag coefficient?

Project

Test the same model in an open jet test section and in a closed jet test section of the same dimensions. Compare the measured data. What can you say about the two sets of data?

References

1. Pope, A., Rae, W. H., and Barlow, J. B., *Low Speed Wind Tunnel Testing*, 3rd ed., John Wiley and Sons, Inc., 1999.
2. Cogotti, A, "Automobile Aerodynamics: An Update, SAE SP706, New Measurement Techniques the Wind Tunnel, 870243," *Autotechnologies Montecarlo* 85, February 1985.
3. Anderson, J. D., *Fundamentals of Aerodynamics*, 2nd ed., McGraw-Hill, 1991.
4. Kelly, K. B., Provencher, L. G., and Shenkel, F. K., "The General Motors Engineering Staff, Aerodynamics Laboratory—A Full-Scale Automotive Wind Tunnel," SAE Paper 82-0371, 1982.
5. Sovran, G., and Klomp, E. D., "Experimentallly Determined Optimum Geometries for Rectilinear Diffusers with Rectangular, Conical or Annular Cross-Section," *Fluid Mechanics of Internal Flows*, Elsevier Publishing Co., 1967, pp. 270–319.
6. Parkinson, G. V., "A Tolerant Wind Tunnel for Industrial Aerodynamics," 5th Colloquium on Industrial Aerodynamics, Aachen, June, 1982.
7. Mikhail, M. N., and Rainbird, W. J., "Optimum Design of Wind Tunnel Contractions," AIAA Paper 78-819, 1978.
8. Vandry, F., and Wieghardt, K., "Experiments on a Slotted Wall Working Section in a Wind Tunnel," Aeronautical Research Council, Current paper 206, London, 1955.
9. Berndt, S. B., and Sorensen, H., "Flow Properties of Slotted Walls for Transonic Test Sections," AGARD Conference Proceedings174, Paper 17, 1975.
10. Raimondo, S., and Clark, P.J.F., "Slotted Wall Test Section for Automotive Test Facilities," AIAA 12th Aerodynamic Testing Conference, Paper AIAA-82-0585-CP, Williamsburg, March, 1982.
11. Flay, R. G. G., Elfstrom, G. M., and Clark, P. J. F., "Slotted Wall Test Section for Automotive Test at Yaw," SAE Paper 830302, Detroit, 1983.

12. Wandby-Smith, P. M., and Rainbird, W. J., "Some Principles of Automotive Aerodynamic Testing in Wind Tunnels with Examples from Slotted Wall Test Section Facilities," SAE Paper 850284, Detroit, 1984.

13. Baals, D. D., and Corliss. W. R., "Wind Tunnels of NASA," NASA SP440, Washington, DC, Government Printing Office, 1981.

14. Dayman, B., "Free-Flight Testing in High-Speed Wind Tunnels," North Atlantic Treaty Organization Advisory Group for Research and Development, AGARDograph 113, May 1966.

15. Goethert, B. H., *Transonic Wind Tunnel Testing*, Pergamon Press, 1961

16. Garner, H. C., Rogers, E. W. E., Acum, W. E. A., and Maskell, E. C., "Subsonic Wind Tunnel Wall Corrections," North Atlantic Treaty Organization Advisory Group for Research and Development, AGARDograph 109, October 1966.

17. Hamsten, B., and Christensen, F. M., "Correlation Tests in a Climatic Wind Tunnel," Automobile Aerodynamics: An Update, SAE SP706 870249, Detroit, 1975.

18. Gallington, R., and Sisson, G., "Flow Visualization Using a Computerized Data Acquisition System," 2nd International Symposium on Flow Visualization, Bochum, September 1980.

19. Fackrell, J. E., and Harvey, J. K., "The Aerodynamics of Isolated Road Wheel," *Proceedings of the 2nd AIAA Symposium on the Aerodynamics of Sports and Competition Automobiles*, Vol. 16, pp. 119–125, 1974.

20. Stapleford, W. R., and Carr, G. W., "Aerodynamic Characteristics of Exposed Rotating Wheels," MIRA Report No. 1970/2, 1970.

21. Wiedemann, J., "The Influence of Ground Simulation and Wheel Rotation on Aerodynamic Drag Optimization—Potential for Reducing Fuel Consumption," SAE paper 960672, 1996.

22. Stapleford, W. R., Carr, G. W., Elofsson, P., and Bannister, M., "Drag Reduction Mechanisms Due to Moving Ground and Wheel Rotation in Passenger Cars," SAE 2002-01-0531, 2002.

23. Stapleford, W. R., and Carr, G. W., "Aerodynamic Characteristics of Exposed Rotating Wheels," MIRA Report No. 1970/2, 1970; Wäschle, A. "The Influence of Rotating Wheels on Vehicle Aerodynamics—Numerical and Experimental Investigations," SAE 2007-01-0107, 2007.

24. Fanger-Vexler, S., Katz, J., and Fux, A., "Full-Scale On-Road Study of the Effect of Automobile Shape on Its Aerodynamic Characteristics and Comparison with Small-Scale Wind Tunnel Results," SAE Paper 85 0287, *SAE Transactions*, Vol. 94, Part 2, 1985.

25. Roussilon, G., "Contribution to Accurate Measurement of Aerodynamic Drag on a Moving Vehicle from Coast-Down Tests and Determination

of Actual Rolling Resistance," *Journal of Wind Engineering and Industrial Aerodynamics*, Vol. 9, 1981, pp. 33–48.

26. Korst, H. H., and White, R. A., "Evaluation of Vehicle Drag Parameters from Coast-Down, Experiments Conducted Under Non-ideal Environmental Conditions," *Journal of Fluid Engineering*, Vol. 103, March 1981, pp. 133–140.
27. Beauvais, F. N., Tignor, S. C., and Turner, T. R., "Problems of Ground Simulation in Automotive Aerodynamics," SAE Paper 68-0121, Detroit, 1968.
28. Walston, W. H., Buckley, F. T., and Marks, C. H., "Test Procedures of the Evaluation of Aerodynamic on Full-Scale Vehicles in Windy Environments," SAE Paper 76-0106, Detroit, 1976.

Chapter 4
Computational Aerodynamics for Ground Vehicles

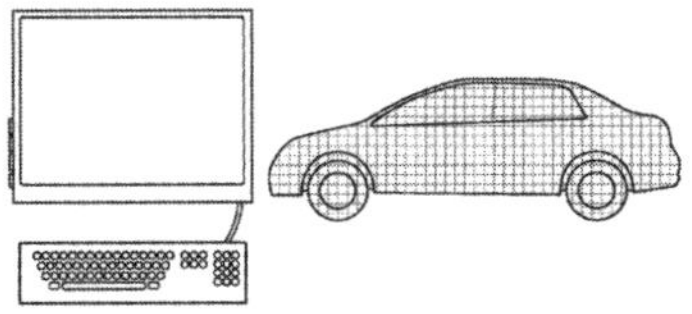

The success of computational physics relies heavily on reliable experimental results. For example, in order to simulate a wind tunnel test, the numerical scheme must be such that it simulates the test environment (i.e., the wind tunnel itself). The computational scheme is validated by comparing the pressure gradient and the boundary layer distribution in the empty test section of the computational model with experimental values. The force and pressure on simplified models are thereafter compared with the values obtained from experiment [4-1]. However, once the experimental results become acceptable data for computation, the computational method is by far the most responsive to various what-if scenarios.

4.1 Making a Case for Computational Fluid Dynamics CFD

The three primary methods to test the theory or the design of a ground vehicle are wind tunnel testing, road test, and computational method. Each of these methods has its benefits and disadvantages, depending on the phase of the design when the test is carried out. Over time, enough information has been accumulated to enable modern designers to embark on a vehicle design without having to start from scratch. The same data have enabled such a tremendous improvement in computational method that the latter has arguably outranked the other approaches as a test tool, meeting the shortcomings of both the wind tunnel and the road test approaches. For example, experimental optimization of cooling air flow in vehicles is difficult and time consuming. Production run vehicles have so many different components and options within a model that detailing the complex geometry to size can be a difficult if not near-impossible task. Fortunately, numerical schemes have been developed that adequately capture the flow characteristics on a vehicle [4-2]. Computational results can be very detailed and precise because there is no problem of component accessibility to contend with. Pressure, velocity, and temperature variables are usually available throughout the model. Computational systems are portable and can usually be expanded to other applications. Table 4.1 elucidates the relative benefits of the computational method.

Table 4.1 Benefits of CFD over Wind Tunnel and Road Test Methods

Relative Shortcomings of the Wind Tunnel Method	Relative Shortcomings of the Road Test Method	Computational Fluid Dynamics (CFD) Benefits
Build cycle—wind tunnel design, build and certification—is long and cumbersome Expensive to build and to operate full-scale wind tunnels Size limitation of test sections Underside wind condition is unrealistic Engine cooling is not considered; passenger compartment flows are not taken into account Interference of probes used in wind tunnel data collection	Rig set-up is a long process and requires continuous adjustment for varying terrains Rig set-up is expensive Cannot capture a range of wind conditions Test and tester are at the mercy of the weather Hard to isolate compartment flows Instrumentation for data collection may interfere with environment and affect data	Reduction in development cycle Relatively economical No size limitation as long as the computing power is not exceeded A dynamic environment is easily simulated Compartments that are hard to reach or to rig for data collection are easily modeled in CFD Interference of data collection materials is eliminated Particularly useful in the early phase of product cycle pretest planning No initial hardware procurement required Quick aerodynamic evaluation of many different shapes Fine-tuning for final selection

4.2 Modeling

Model preparation is perhaps the most labor intensive of computational methods. Although different packages have been developed to clean-up a computer-aided design CAD package, the best solid model is a clean-from-scratch solid model. To appropriately model the flow around a vehicle, it is important to be aware of the features of flow around the vehicle. The general features are often captured correctly; the challenge is the adequate representation of specific features such as singularity points, steep transition points, and points of inflexion on the vehicle surface. In the big picture of the vehicle profile, these points seem inconsequential, but failure to capture them in aerodynamic modeling has a significant consequence in the validity of the computational result.

The actual flow around a vehicle is three-dimensional, with steep pressure gradients. There are two main regions of separated flow: open regions and closed regions. The open regions of separated flow consist of the A-post, the wheels, vehicle rear end, and vehicle underside. The closed regions of separated flow consist of such components as mirrors, door handles, headlights, windshield wipers, and hood insignia.

For a CFD code to be an effective modeling tool, the code benchmark result must:

Correlate with validated wind tunnel test results;

Take a shorter time than wind tunnel test or road test;

Cost less than wind tunnel test or road test; and

Address details (answer "why," i.e., explain local occurrences).

4.3 Methods of Modeling

Flows are generally modeled as linear or nonlinear. In linear models, the fluid is treated as ideal. Hence we have a non-viscous, incompressible, irrotational flow. The flow field is conservative, providing a solution to Laplace equation [4-3, 4-4]. The vortex lattice method (VLM) is a linear method. Linear methods are independent of Reynolds number.

In nonlinear methods, the flow can be inviscid (Euler), time-averaged viscous (Reynolds averaged Navier-Stokes—RANS), or viscous, unsteady (large eddy simulation—LES, direct numerical simulation—DNS). The peculiarities of each of the four types are listed here:

Euler Flow
• non-viscous • does not account for Re (inertia force/viscous force) • inadequate in most flow conditions • minimal CPU time • provides general trend
Reynolds Averaged Navier-Stokes (RANS)
• uses turbulence modeling • includes viscous term • can determine pressure and friction drag

(continues)

Large Eddy Simulation (LES)
• real flow behavior
• incorporates large eddies and small eddies
• simulates large eddies using Navier-Stokes (N-S) equations
• small eddies are modeled
• assumed general behavior of flow is independent of dominant flow
Direct Numerical Simulation (DNS)
• straightforward solution of N-S equations
• no turbulence modeling
• demanding computational space
• can presently only afford for 1-D simulation and simple configurations

4.4 Meshing

After model preparation, the next practically difficult task is meshing. The quality of the mesh directly influences the result of an analysis. Ideally, a small grid size is desired since the smaller the mesh, the better the computational model represents the real object (Fig. 4.1). Unfortunately, the computational demand increases (exponentially) as the grid size is reduced. The goal is therefore to optimize the grid sizing, apportioning smaller grid sizes to critical areas, and maximizing the grid sizes in smooth, relatively panel surfaces.

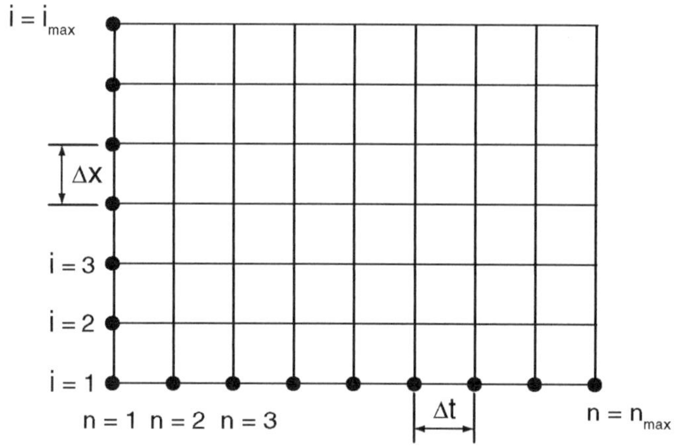

Fig. 4.1 Grid set-up

Unstructured grids are preferred in modeling an object with varying surfaces because they capture the surface better than structured grids. But structured grids are, as expected, easier to construct. With the advancement in software, the relative time demand in building unstructured grids is substantially reduced, but computer memory demand for unstructured grids still surpasses that of the structured grid by far.

Because of the rapid changes in the profile of a car, unstructured grids are recommended for modeling cars. A car can be considered as consisting of regions of rapid changes in flow, wake regions, and high curvature regions. The closed regions—mirrors, door handles, headlights, windshield wipers, and hood insignia—are typical regions of rapid changes in flow, and the open regions, the rear end in particular, are the wake regions.

Buses and trailers, however, have a continuous, near-flat body that the less-expensive structured grid can take advantage of.

4.5 Boundary Conditions

The boundary conditions stipulated have a strong influence on the accuracy of the computation. The methods of the application of the boundary conditions are significant to the convergence properties of the solution.

Boundary conditions can be explicit or implicit (Fig. 4.2). Explicit boundary conditions are generally in the source of the formulated matrix, and implicit boundary conditions are in the matrix, giving a more compact arrangement. Large source terms typically result in a slow convergence. The convergence can be accelerated by the use of multistage or iterative time integration methods (Fig. 4.3). Whereas explicit schemes are only conditionally stable, implicit formulations are always stable. This benefit, however, does not necessarily make the implicit method more appealing, since it generally requires a greater CPU allocation. Whether the implicit method or the explicit one should be used depends on the goal of the computation. If the transient behavior of the flow is of interest, the explicit method is more likely to give the desired picture because it is time-stepped. But if only the steady-state results are of interest, the implicit method is generally a better choice.

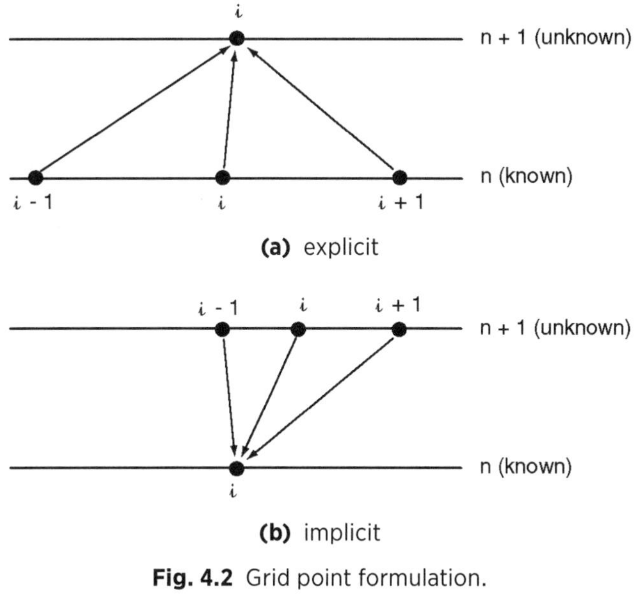

Fig. 4.2 Grid point formulation.

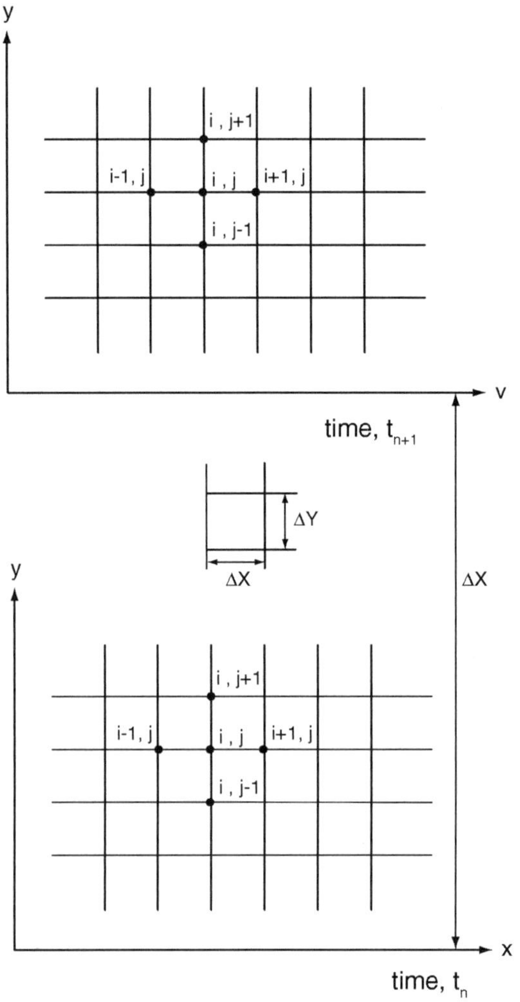

Fig. 4.3 Typical computational grid for a source flow problem

Other types of boundary conditions are addressed below.

4.5.1 Undefined (Unspecified) Boundary Condition

In undefined or unspecified boundary conditions, the boundary conditions are, by default, prescribed by averaging the results from local solution points. Undefined boundary specifications are, in general, not advisable as they may fail to capture the physics of the flow at the boundary.

4.5.2 Frozen Boundary Condition

The frozen boundary condition is arguably the opposite of the undefined boundary condition. The undefined boundary condition is a sort of whatever-you-get, but the frozen boundary condition, as the name implies, stays fixed through all iterations. An example would be a constant temperature at a surface, irrespective of the surrounding dynamics. Usually, a source function is required to ensure the physical correctness of a frozen boundary.

4.5.3 Reflection Boundary Condition

The reflection boundary condition is a solid, slip wall boundary condition. It simulates a symmetric plane. Reflection boundary condition is useful in the reduction of memory and computational space. A representative section of the entire physical model is analyzed, and the result is applied to the entire structure. The reflection boundary condition can only be taken advantage of in symmetrical bodies within the same environment.

4.5.4 Free Stream Boundary Condition

The free stream boundary condition is employed in flow-free stream outer boundaries. The flow direction at the boundary and the free stream conditions are used to determine the flow field boundary condition.

4.5.5 Inflow Boundary Condition

The inflow boundary condition indicates the condition of the flow entering the system. It is relatively an independent variable. The inflow boundary condition is perhaps the easiest to impose on a flow because it is often a given, directly or indirectly.

4.5.6 Outflow Boundary Condition

The outflow boundary conditions are employed at the exit boundary of internal flows, as at a diffuser exit. They are also used at the downstream outflow regions of external flows. A mass flow or back pressure specification are examples of outflow boundary conditions. Outflow boundary conditions are often obtained from an earlier experiment on the physical model.

4.5.7 Viscous Wall Boundary Condition

The viscous wall boundary condition is a commonly imposed condition. It imposes a no-slip flow condition ($\frac{\partial u}{dy} = 0$) and a zero pressure gradient. The no-slip condition implies that at a solid boundary, the fluid has zero velocity relative to the boundary. Although the no-slip condition is commonly used in viscous flow modeling, the no-penetration boundary condition is sometimes more representative of the physics of the flow. In the no-penetration condition, the fluid velocity normal to the wall is set equal to the wall velocity in that direction, whereas the fluid velocity parallel to the wall is unrestricted. For example, if the boundary layer effect is negligible (as in momentum-dominant flow), the fluid velocity is zero in a direction normal to a wall that is stationary in the direction of the normal fluid flow; the wall may move tangentially to the fluid.

4.5.8 Non-viscous Wall Boundary Condition

The non-viscous wall boundary condition forces a flow tangency at the wall while the same flow velocity is assumed at the flow adjacent to the boundary. Typically, the presumed pressure at the boundary is computed via an interpolation of interior pressures. A first-order extrapolation is usually prescribed over a zero-order extrapolation for this type of boundary condition, as the former depicts continuity better, especially when the pressure does not vary substantially with the normal boundary. Non-viscous boundary condition has the disadvantage of causing a mere numerical (nonphysical) pressure when extrapolations are performed across discontinuities.

4.5.9 Bleed Boundary Condition

In the bleed boundary condition, mass flow is allowed through a porous, viscous wall. Mass flows out (bleeds out) of the computational region or flows in (blows in) to the computational region through the boundary.

4.5.10 Coupled Boundary Condition

Coupled boundary conditions are implemented between zones that merge. The coupling ensures a seamless transition from one region of the coupling to the other. The coupling process employs geometric interpolation between the interior domains of each domain to generate the boundary data, which is then used as the boundary condition for the computational cycle in each domain.

Coupled boundaries should not be placed in regions of large flow gradients or in shock flow regions. Orthogonal flow regions and large changes in grid stretching should also be avoided.

4.6 Solution Methods

After the problem has been formulated, the mesh established, and the boundary conditions imposed, the problem is solved using a solution method that is compatible with the problem formulation. The solution options include the panel method, Euler method, RANS method, LES method, and DNS method.

4.6.1 Panel Method

A potential flow model, the panel method is a linear, simplified method of solution of flow problems. The flow is assumed non-viscous and incompressible. In essence, the panel method solves Laplace equations. In the panel method, the velocity potential at any point in the fluid is represented by surface distributions of singularities over the boundary surface. This leads to an integral equation that is then solved for the source strength or the velocity potential [4-5, 4-6, 4-7]. From the latter, the streamlines and such properties as the velocity, pressure drag, and lift force are determined.

The panel method can give dependable results for wide uniform surfaces as found with buses, trucks, and trains. For rapidly changing contours as found in cars, it still gives a trend and useful first pass information that can serve as input for advanced analysis in which viscous effects and flow separation are accounted for. The panel method is especially good for high Reynolds number flows (Re>105) that are fully attached, since viscous effects are practically negligible in such flows. Being 2-D due to irrotationality, potential flows take minimum computer memory, only providing surface flow results.

4.6.2 Euler Method

When the viscosity terms are termed negligible, we have the reduced form of N-S referred to as Euler equations. A first-order numerical method, the Euler method uses an explicit approach to provide the numerical integration of ordinary differential equations with the initial values given. A little more advanced than the panel method, it incorporates vorticity and simulates flow separation. Although useful in high-speed flows (where inertia dominates), the results of Euler equations fail to capture the effects of turbulence or of flow separations, seeing these are effected by viscosity.

4.6.3 Reynolds Averaged Navier-Stokes (RANS)

The RANS method approaches the reality of the physical flow more than the Euler or the panel method. Fluid viscosity is taken into account and the flow is generally treated as a turbulent flow, whereby both steady and unsteady components are accounted for. RANS flows are time dependent, and the flow is averaged over time [4-8].

4.6.4 Large Eddy Simulation (LES)

In LES, the large scales are modeled directly and a turbulence model is introduced for the small scales. Although not as computer-power demanding as DNS, it is still a challenge in spite of the advances in computers. The 2-D modeling has generally been more successful than the 3-D because the former is less demanding of computing power. A number of 3-D analyses that successfully implement the LES solution method are often a uniform extension of the 2-D—rather simplistic when compared with the 3-D physical model.

4.6.5 Direct Numerical Simulation (DNS)

Perhaps the most representative of the physics of the flow, the DNS approach makes no assumption or approximation about the flow. There is no turbulence modeling. Rather, the turbulence is computed directly. DNS is an ideal computational approach and, at present, a lofty benchmark due to the massive (and yet unrealized) memory requirement for a typical engineering model; the extensive grid requirement; and the lengthy computation time, all of which tremendously add to cost. The DNS is at present a useful educational tool and a valuable benchmark for a 1-D solution of any of the other methods. Table 4.2 itemizes the properties of the solution methods.

Table 4.2 Peculiarities of Main Solution Methods in CFD

Panel	Euler	Reynolds Averaged Navier-Stokes (RANS)	Large Eddy Simulation (LES)	Direct Numerical Simulation (DNS)
Solves Laplace equation Linear Flow is ideal Good for large vehicles (e.g., buses) Panel singularities Sources and sinks are on panel surface Dipoles on wake surfaces No viscosity Cannot simulate flow separation	Vorticity Flow separation No viscosity	Turbulent flow Models viscosity Steady, unsteady components Averaged flow (time dependent)	Large scale eddies are computed directly Small scale eddies are modeled	No turbulence modeling needed Turbulence computed directly High cost Computation time Grid requirement

Introduction of fudge factors, model modification, and mesh refinement are among the steps taken to improve the result of a CFD run. The standard $k - e$ model is perhaps the best understood and the most studied turbulence model and is typically used in RAN and LES solution methods [4-9, 4-10].

Fudge factors can cause over prediction or under prediction of drag or pressure. Steady-state results are, in general, not affected by the choice of fudge factor.

Mesh refinement can show an increase in the size of, say, the recirculation bubble behind the bumper or on the hood of a vehicle. Mesh refinement is an economic way to capture the details in a particular region on the vehicle. The region of interest can be substructured and densely meshed to show results that may have escaped the first pass in which the mesh is relatively sparse. Often, the turbulence model fails to capture a primary vortex in the solution. A further mesh refinement helps to reveal this and properly captures the vortex [4-11].

4.7 Interpretation of Results

After the solution of a flow problem is the post-processing phase. In this, plots are typically generated to provide a visual capture of the computed results to compare the numerical solution with theory (if a theoretical solution exists) or with the physics of the problem (i.e., real-life findings). The output data that are selected to be plotted depend on the goal of the computation and, certainly, on the customer request (Fig. 4.4).

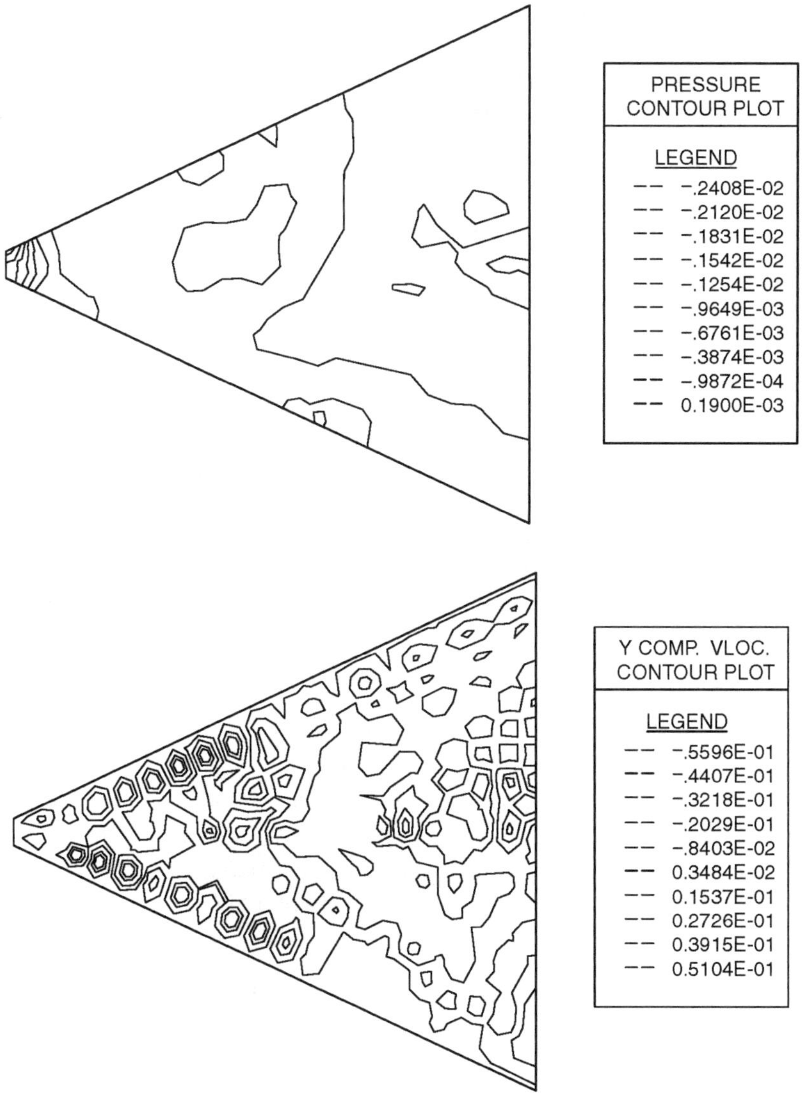

Fig. 4.4 Plot of pressure and velocity distribution in a diffuser

The areas around the bluff body and in the wake of a vehicle strongly affect the force and the surface pressure on a vehicle. Therefore, one of the significant validators of a numerical scheme is the degree of correlation between the computational and experimental pressure readings in these regions. Further refinement of the primary solution can be performed via submodeling. In submodeling, critical areas are further fine-meshed to provide more accurate results in the next batch of computation.

Computational results serve to validate and are validated as well by experimental and theoretical results. In this, there is an added benefit of the validation of tests that have been approached by methods other than computational. In general, when analyzing vehicle CFD results, the decrease of speed indicates a pressure field's sensitivity to separation along the aft-end of a vehicle. There is less energy in the boundary layer to negotiate the adverse curvature of the vehicle.

At low Re, usually corresponding to low speeds, the impact of Re effects for scale model and full-scale CFD runs is small. It is therefore recommended to run a scale model for low speeds because it is less expensive and less demanding on computer memory. The scale model may, however, fail to adequately capture trailing edge vortices because of relatively lower vortex strength from the scale model vortices compared with that of full-scale vortices. Up to 35% drag reduction can be achieved over an already streamlined vehicle by blowing control of aft-flow separation and vortex-induced drag reduction. It is therefore important to capture the vortex strength, as this contributes to a more accurate result in which there is a higher drag reduction [4-12]. At higher speeds (>65 mph [105 km/h]), the full-scale Re gives a more accurate result. Also, at higher speeds there is a quicker pressure recovery from the stagnation point at the forward edge of the vehicle.

4.8 CFD Packages

There are several commercial packages for CFD, but they are usually expensive and (for commercial reasons) tightly controlled. The consequences of these concerns are, respectively, limited accessibility due to cost and difficulty in adapting a package to a situation or design that the package has not properly addressed. Sometimes, the analyst writes a code of a few lines—a patch—to interface with the commercial code in order to address the problem of interest.

Because it is almost impossible to find a code that fully meets every customer's requirement, many major companies and research centers write their own in-house codes to solve their particular problems [4-13]. Sometimes the in-house code complements a commercial code, and sometimes it is a standalone, completely replacing a commercial code. Ironically, a common, yet unintended outcome of in-house codes is that they become so big and versatile due to add-ons that they eventually evolve into commercial codes.

The evolution of computational methods has reached a level that the correction factor can be fudged into a numerical solution, given the model reduction scale factor. Although there is still work to be done in reducing the model preparation time, time is already being saved in studying various what-if situations without building physical models to test, and without the waiting time for or the direct cost of using wind tunnels.

4.9 Turbulence Modeling

Equations are simplified by making some generally acceptable assumptions and modifying the generalized Navier-Stokes equations based on the assumptions. One such assumption is the modeling of turbulence rather than utilizing the full, physical turbulence. A turbulence model only accounts for the pronounced average effects of turbulence, effects such as momentum transport and energy dissipation. The RANS and LES methods, both discussed earlier, are examples of turbulence modeling techniques.

Summary

Computational method is the most robust of the three methods used in testing vehicle design; the other two are wind tunnel testing and road testing. Benefits of computational testing over the other methods include the convenience of making changes on the fly, thereby allowing for quick aerodynamic evaluation of many different shapes, the ability to model compartments that are hard to reach, the absence of size limitation (assuming uninhibited computation power), and the ease of fine-tuning for the final selection of components.

For a CFD code to be acceptable as an effective modeling tool, the code benchmark result must correlate with existing wind tunnel results; it must produce the result considerably faster than a wind tunnel test or road test; it must cost less than the other methods; and it must be able to address the effect of component changes.

Stages in computational approach to design and analysis are model selection, meshing, boundary condition specification, solution, and interpretation of results.

Key Terms

Euler flow: Flow is non-viscous; does not account for Reynolds number.

Ideal flow: Flow is non-viscous, incompressible, and irrotational; ideal flow treatment is limited to 2-D flows.

Linear model: Model in which the flow is treated as ideal.

Mesh: The division of a surface or structure into cells or partitions for computational purposes.

Review Questions

1. List three benefits of computational method over road test.
2. List three benefits of computational method over wind tunnel testing.
3. List three requirements of a CFD package for the package to be considered an effective tool.
4. What are the benefits of Euler flow modeling over Reynolds Averaged Navier-Stokes (RANS) modeling? What are the shortcomings?

Application Question

Direct Numerical Simulation (DNS) method is perhaps the most appealing computational approach, but because of the large computational memory that it requires, it is hardly used for flows other than one-dimensional. Given the current memory limitation, how would you piecemeal a full three-dimensional model using DNS?

Exercise

Consider a sports car traveling at 300 kilometers per hour on an open, straight and level highway. Model the sports car using the dimensions provided as guidelines for the geometry. Study the flow on the car and measure the aerodynamically induced forces on the vehicle surfaces. Justify your boundary conditions and your choice of solution method. (Length = 4.5 m, Height = 1.5 m, width = 2 m. Vehicle net dry *weight* = 700 kg)

Project

Develop a linear model for the flow along the side of a 53-ft. truck trailer. Also develop a Euler model, an LES model, and a DNS model. Compare the drag obtained in each of the models. What do you observe? What do you conclude about model selection for flow on a large, unencumbered surface? Based on your conclusion, which of the methods would you consider optimum for the flow along the fuselage of a large airplane?

References

1. Fischer, O., Kuthada, T., Widdecke, N., and Wiedemann, J., "CFD Investigation of Wind Tunnel Interference Effects," SAE Paper 2007-01-1045, 2007, doi:10.4271/2007-01-1045.
2. Baeder, D., Indinger, T., and Adams, N., "Comparison of Numerical Simulations with Experiments of Bluff Bodies Including Under-hood Flow," SAE Paper 2011-01-0171, 2011.
3. NASA, "Vortex-Lattice Utilization," NASA SP-405, NASA-Langley, Washington, 1976.
4. Katz, J., and Plotkin, A., *Low-Speed Aerodynamics*, 2nd ed., Cambridge University Press, 2001.
5. Lamb, H., *Hydrodynamics*, 6th ed., Cambridge University Press, 1932.
6. Linton, C. M., and McIver, P., *Handbook of Mathematical Techniques for Wave/ Structure Interactions*, Chapman & Hall/CRC, 2001.
7. Newman, J. N., *Marine Hydrodynamics*, MIT Press, 1977.
8. Rhee, S. H., and Stern, F., "Unsteady RANS Method for Surface Ship Boundary Layer and Wake and Wave Field," *International Journal for Numerical Methods in Fluids*, 2001;37, pp. 445–478.
9. Haroutunian, V., and Engelman, M. S., "On Modeling Wall-Bound Turbulent Flows Using Specialized Near-Wall Finite Elements and the Standard $k - e$ Turbulence Model," Advances in Numerical Simulation of Turbulent Flows, ASME 1991.
10. Haroutunian, V., and Engelman, M. S., "Two-Equation Simulation of Turbulent Flows: A Commentary on the Physical and Numerical Aspects," Advances in Finite Element Analysis in Fluid Dynamics, ASME 1993.
11. Hajiloo, A., Williams, J., Hackett, J. E., and Thompson, S. A., "Limited Mesh Refinement Study of the Aerodynamic Flow Field Around Car-like Shape: Computational versus Experimental Fluid Dynamics," Vehicle Aerodynamics: Wind Tunnels, CFD, Aeroacoustics, and Ground Transportation Systems, SP-1145, 960677, SAE, 1996, pp. 63–73.

12. Englar, R. J., Smith, M. J., Niebur, C. S., and Gregory, S. D., "Development of Pneumatic Aerodynamic Concepts for Control of Lift, Drag and Moment plus Lateral/Directional Stability of Automotive Vehicles," Vehicle Aerodynamics: Wind Tunnels, CFD, AeroAcoustics, and Ground Transportation Systems, SP-1145, SAE, February 1966, pp. 27–38.

13. Hucko, W. H., and Sovran, G., "Aerodynamics of Road Vehicles," *Annual Review of Fluid Mechanics*, 1993, No. 25, pp. 485–537.

Chapter 5
Vehicle Stability and Performance

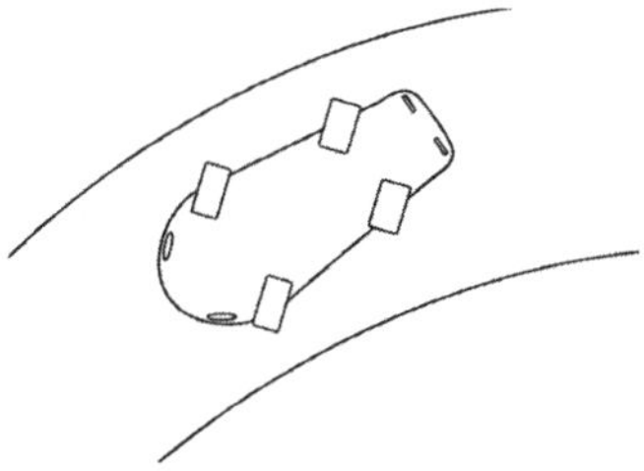

Analysis of the air flow around a motor vehicle and the relationship between its shape and the resulting aerodynamic characteristics helps us to refine mission specification and to design for aerodynamic competence. Aerodynamic challenges must be given proper attention at the primary stages of vehicle design. Although a primary goal is the creation of a low drag shape, that will not suffice until we have taken into consideration other concerns, such as stability and performance.

The first phase of automobile performance is the traction-limited phase. This is when there is a high power-to-weight ratio or short run. In this phase, the aerodynamic downforce improves the performance, with the drag coefficient remaining constant. Aerodynamic devices produce performance improvement only for short runs with high power-to-weight ratio. As the runs increase or the power-to-weight ratio reduces (or both), the vehicle still experiences an improvement in performance, but the induced drag also increases to the point that it outweighs the gains in the downforce. In the long runs with low power-to-weight ratio, the aerodynamic lift produces an improved performance. But, as in the case of moderate run, the resulting induced drag causes an overall reduction in performance [5-1].

5.1 Traction

The functional characteristics of a ground vehicle are generally described in terms of its performance, which in essence consists of its ability to accelerate and decelerate quickly, its torque for climb or for load pull, and its ride. The ride describes the combination of passenger comfort and handling or steering response. The performance of a vehicle depends in great part on the tractive-ability of the tires and on the drag forces experienced by the vehicle. The loss of the coefficient of road adhesion, the normal load on the drive axles, and the power plant and transmission contributions must be accounted for in order to fully analyze vehicle performance. This subject is well treated in the literature [5-2, 5-3, 5-4]. A simplified method to determine the maximum tractive force, F_{max}, for vehicle tire-ground contact follows. The presentation in this book takes after the approach of Wong [5-4].

5.1.1 Passenger Car

In the longitudinal direction, the forces acting on the car (Fig. 5.1) are aerodynamic resistance R_a, the rolling resistance of the front and rear tires R_{rf}, R_{rr}, respectively, grade resistance R_g (= $W \sin \theta_s$ for a car of weight W), drawbar load R_d, and the tractive effort at the tires F_f, F_r. The equation of motion for this arrangement is

$$m\frac{d^2x}{dt^2} = \frac{W}{g}a = F_f + F_r - (R_a + R_{rf} + R_{rr} + R_d + R_g) \tag{5.1}$$

Rearranging

$$F_f + F_r = \frac{W}{g}a + R_a + R_{rf} + R_{rr} + R_d + R_g \tag{5.2a}$$

Grouping the forces, we can replace $F_f + F_r$ with ΣF and $R_{rf} + R_{rr}$ with ΣR_r. When substituted in (5.2a) we obtain

$$\Sigma F = \Sigma R_r + R_a + R_d + R_g + (W/g)a \tag{5.2b}$$

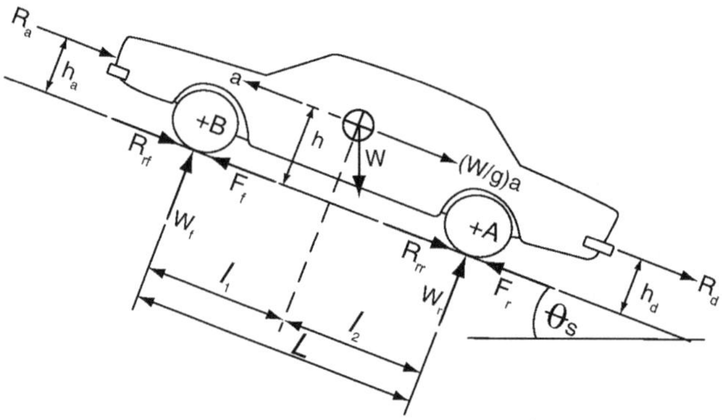

Fig. 5.1 Forces influencing the performance of a car

Taking moments about point A where the rear tires contact the ground, we obtain the normal load, W_f, on the front axle.

$$W_f = \frac{Wl_2 \cos\theta_s - R_a h_a - h(W/g)a - R_d h_d - Wh\sin\theta_s}{L} \tag{5.3}$$

where

l_2 is the distance between the rear axle and the car center of gravity;
h_a is the height of the point of application of the aerodynamic resistance (drag);
h is the height of the center of gravity (c.g.);
h_d is the height of the drawbar hitch;
L is the wheelbase; and
θ_s is the terrain slope.

Similarly, the normal load, W_r, on the rear axle is obtained by taking moments about point B. The result is

$$W_r = \frac{Wl_1 \cos\theta_s + R_a h_a + h(W/g)a + R_d h_d + Wh\sin\theta_s}{L} \tag{5.4}$$

For passenger cars, h_a; h_d; h. Therefore Eqs. 5.3 and 5.4 practically reduce to Eqs. 5.5 and 5.6, respectively.

$$W_f = \frac{Wl_2 \cos\theta_s}{L} - \frac{h}{L}(R_a + aW/g + R_d + W\sin\theta_s) \tag{5.5}$$

$$W_r = \frac{Wl_1 \cos\theta_s}{L} + \frac{h}{L}(R_a + aW/g + R_d + W\sin\theta_s) \tag{5.6}$$

When Eq. 5.2b is substituted into Eq. 5.5, the normal load on the front axle takes the form:

$$W_f = \frac{Wl_2 \cos\theta_s}{L} - \frac{h}{L}(\Sigma F - \Sigma R_r) \tag{5.7}$$

The first term of the definition is the static load (i.e., the load on the vehicle at rest). The second term is the dynamic load transfer.

The normal load at the rear axle is similarly expressed as

$$W_r = \frac{Wl_1 \cos\theta_s}{L} + \frac{h}{L}(\Sigma F - \Sigma R_r) \tag{5.8}$$

If we denote the coefficient of road adhesion (i.e., the adhesion of the road to the tire) by μ, then the maximum tractive force, F_{max}, for the front-wheel drive can be written as

$$F_{max} = \mu\left[\frac{Wl_2 \cos\theta_s}{L} - \frac{h}{L}(\Sigma F - \Sigma R_r)\right] \tag{5.9a}$$

For the rear-wheel drive, the maximum tractive force is

$$F_{\max} = \mu\left[\frac{Wl_1 \cos\theta_s}{L} + \frac{h}{L}(\Sigma F - \Sigma R_r)\right] \tag{5.9b}$$

Since $F_{\max} = (\Sigma F)_{\max}$, we can substitute $F_{\max}$ for ΣF in Eq. 5.9a to obtain

$$F_{\max} = \mu\left[\frac{Wl_2 \cos\theta_s}{L} - \frac{h}{L}(F_{\max} - \Sigma R_r)\right]$$

Collecting like terms and expressing ΣR_r as $f_r W$ (where f_r is the coefficient of rolling resistance) the above expression becomes

$$F_{\max} + \frac{\mu h}{L} F_{\max} = \frac{\mu W}{L}\left[l_2 \cos\theta_s + f_r h\right], \text{ yielding}$$

$$F_{\max} = \frac{\frac{\mu W}{L}\left[l_2 \cos\theta_s + f_r h\right]}{1 + \frac{\mu h}{L}} \tag{5.10a}$$

Similarly, the maximum tractive force for rear-wheel drive is obtained as

$$F_{\max} = \frac{\frac{\mu W}{L}\left[l_1 \cos\theta_s - f_r h\right]}{1 - \frac{\mu h}{L}} \tag{5.10b}$$

Next, we consider the tractor-trailer arrangement and determine the tractive force for the couple.

5.1.2 Tractor-Trailer

In order to determination of the tractive force for a tractor-trailer, the tractor and trailer are taken as separate bodies (Fig. 5.2).

The equation of motion for the tractor-trailer arrangement is

$$m\frac{d^2x}{dt^2} = \frac{W_1 + W_2}{g}a = F_f + F_{r1} + F_{rt} - (R_{a1} + R_{a2} + R_{rf} + R_{rr} + R_{rt} + R_d + R_g) \tag{5.11}$$

where F_{rt} and R_{rt} are, respectively, the tractive force and the rolling resistance at the rear wheel of the trailer—tractor-trailers are generally rear-wheel drive vehicles.

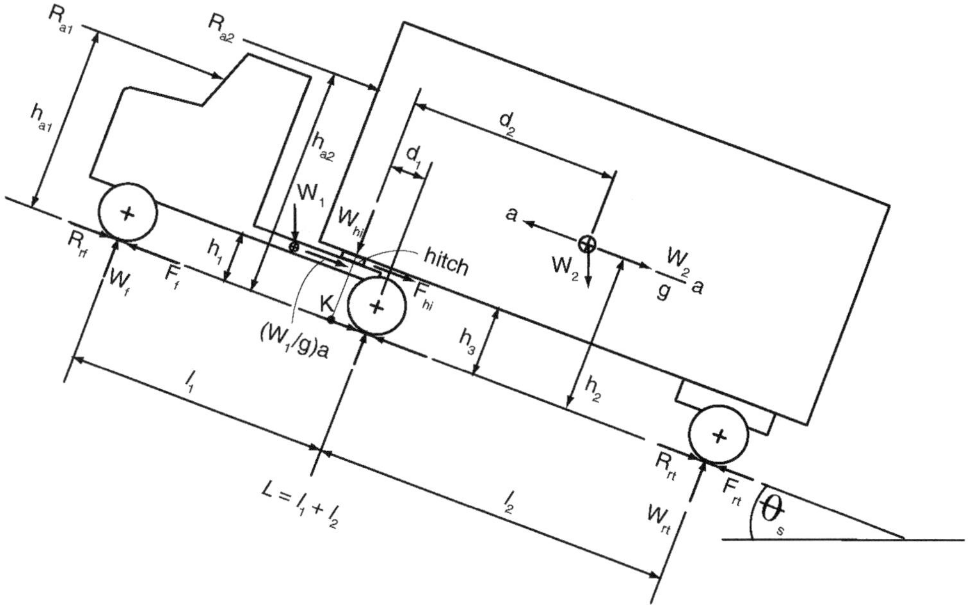

Fig. 5.2 Forces influencing the performance of a tractor-trailer

If F denotes the sum of the tractive forces, and $f_r(W_1 + W_{hi} + W_2)$ replaces three rolling resistances, we can write

$$F = R_{a1} + R_{a2} + f_r(W_1 + W_{hi} + W_2) + R_d + (W_1 + W_2)\sin\theta_s + \frac{W_1 + W_2}{g}a \tag{5.12}$$

where the grade resistance, R_g, has been replaced by $(W_1 + W_2)\sin\,\theta_s$.

Consider just the tractor.

Taking moments about the front wheel-ground contact

$$W_{r1} = \frac{W_1\cos\theta_s l_1 + R_{a1}h_{a1} + \frac{W_1}{g}ah_1 + W_1\sin\theta_s h_1 + F_{hi}h_3 + W_{hi}(l_1 - d_1)}{l_1} \tag{5.13}$$

where

R_{a1} is the aerodynamic resistance of the tractor;
h_{a1} is the height of R_{a1}; and
W_1 is the weight of the tractor.

Applying the approximation $h_{a1};\ h_3;\ h_1$

$$W_{r1} = \frac{W_1 \cos\theta_s l_1 + h_1\left(R_{a1} + \dfrac{W_1}{g}a + W_1 \sin\theta_s + F_{hi}\right) + W_{hi}(l_1 - d_1)}{l_1} \tag{5.14}$$

Since trucks are generally rear-wheel drive, we can write

$$F_{\max} = \mu W_{r1}.$$

But

$$F_{\max} = R_{a1} + \frac{W_1}{g}a + W_1 \sin\theta_s + R_r + F_{hi} \tag{5.15}$$

where the last but one term is the summed rolling resistance ($R_{rf} + R_{rr} + R_{hi}$).

Substituting Eq. 5.15 for $F_{\max}$ in Eq. 5.14,

$$W_{r1} = \frac{W_1 \cos\theta_s l_1 + h_1(F_{\max} - R_r) + W_{hi}(l_1 - d_1)}{l_1}. \tag{5.16}$$

Now, replacing R_r by $f_r(W_1\cos\theta + W_{hi})$, where f_r is the rolling friction coefficient, Eq. 5.16 becomes

$$W_{r1} = \frac{1}{l_1}\left[W_1 \cos\theta_s l_1 + h_1(F_{\max} - f_r(W_1 \cos\theta_s + W_{hi})) + W_{hi}(l_1 - d_1)\right].$$

Now, we can write $F_{\max} = \mu W_{r1}$ as

$$F_{\max} = \mu W_{r1} = \frac{\mu}{l_1}\left[W_1 \cos\theta_s l_1 + h_1(F_{\max} - f_r(W_1 \cos\theta_s + W_{hi})) + W_{hi}(l_1 - d_1)\right].$$

Rearranging the last expression, and noting that for a mild slope, $\cos\theta_s \approx 1$

$$F_{\max} = \frac{\mu}{l_1}\left[\frac{W_1 \cos\theta_s l_1 - h_1 f_r(W_1 + W_{hi}) + W_{hi}(l_1 - d_1)}{1 - \dfrac{\mu h_1}{l_1}}\right] \tag{5.17}$$

The reaction, W_{rt}, at the trailer rear axle is of interest. First, let's establish clarity about the force components at the hitch. When treating the trailer, the components, F_{hi} and W_{hi}, at the hitch must each have a sense opposite to that prescribed for the tractor. This is shown in the free body diagrams at the hitch (Fig. 5.3).

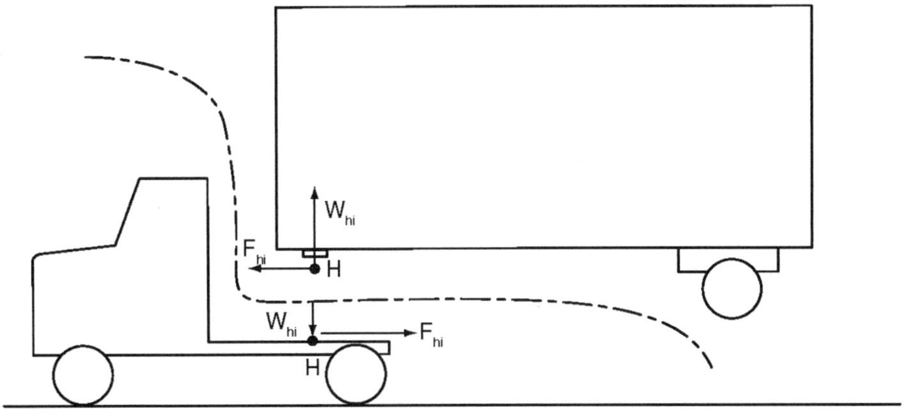

Fig. 5.3 Free body diagram of tractor and trailer at the hitch

Taking moments of the trailer forces about point K (Fig. 5.2) in the line of action of the hitch, and solving for W_{rt}, we get

$$W_{rt} = \frac{W_2 \cos\theta_s d_2 + R_{a2}h_{a2} + h_2 a\frac{W_2}{g} + W_2 \sin\theta_s h_2 + R_d h_2 - F_{hi}h_3}{l_2}.$$

Assume $h_{a2} = h_3 = h_2$. Then the last equation reduces to

$$W_{rt} = \frac{d_2}{l_2}W_2 \cos\theta_s + \frac{h_2}{l_2}\left(R_{a2} + a\frac{W_2}{g} + W_2 \sin\theta_s + R_d - F_{hi}\right) \tag{5.18}$$

For the trailer, the longitudinal force at the hitch is

$$F_{hi} = R_{a2} + \frac{W_2}{g}a + W_2 \sin\theta_s + f_r W_{rt} + R_d \tag{5.19}$$

Substituting Eq. 5.19 into Eq. 5.18 yields

$$W_{rt} = \frac{d_2}{l_2}W_2 \cos\theta_s - \frac{h_2}{l_2} f_r W_{rt}$$

$$W_{rt}\left(1 + \frac{h_2}{l_2} f_r\right) = \frac{d_2}{l_2}W_2 \cos\theta_s$$

$$W_{rt} = \frac{d_2 W_2 \cos\theta_s}{l_2 + h_2 f_r} \tag{5.20}$$

Knowing W_{rt} is critical to the proper sizing of the tires at the trailer rear axle.

5.2 Power Requirement

Assuming equilibrium of forces and steady speed, the power required for the operation of a vehicle can be written as the sum of the power to overcome the aerodynamic drag and that for traction. If *P* is the power supplied by an engine of efficiency, η, the effective power supply is ηP.

$$\eta P = \frac{1}{2}\rho v^3 C_D A + Fv \tag{5.21}$$

As Eq. 5.21 shows, whereas the mechanical power is proportional to the speed, v, the aerodynamic power is proportional to v^3. The relationship is illustrated by Fig. 5.4.

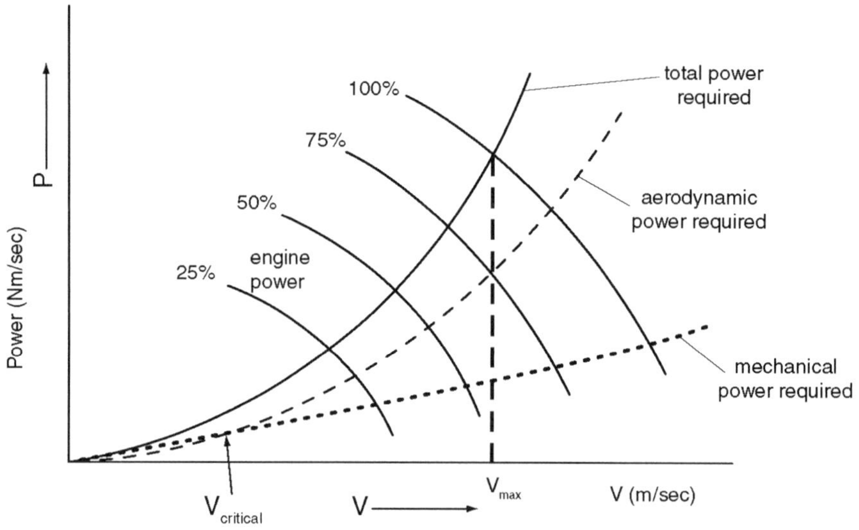

Fig. 5.4 Comparison of mechanical and aerodynamic power requirements

Most of the initial power input in a vehicle is used to overcome the tractive obstacle since the aerodynamic resistance is relatively insignificant until a speed of about 50 mph (80 km/h), after which the effort required to overcome the aerodynamic drag clearly dominates. The intersection of the power required and effective power supplied curves is the equilibrium condition for a given vehicle speed.

Below v critical, the mechanical resistance to rolling dominates. But after that point, the aerodynamic drag becomes dominant because the mechanical resistance merely varies linearly with the speed (Fig. 5.5). It is therefore important to minimize the aerodynamic drag of a vehicle. It is, however, just as important to do so without reducing the stability of the vehicle.

In general, at speeds below 65 mph (104 km/h) on a level road, the rolling resistance is greater than the air drag. As the speed increases, the rolling resistance changes very minimally, but the aerodynamic drag increases and dominates the tire contribution, accounting for up to 30 % of the total resistance for a 30-ton tractor-trailer combination. Therefore, in city driving and short commutes the contribution of the tire to drag is significant but much less so in highway driving.

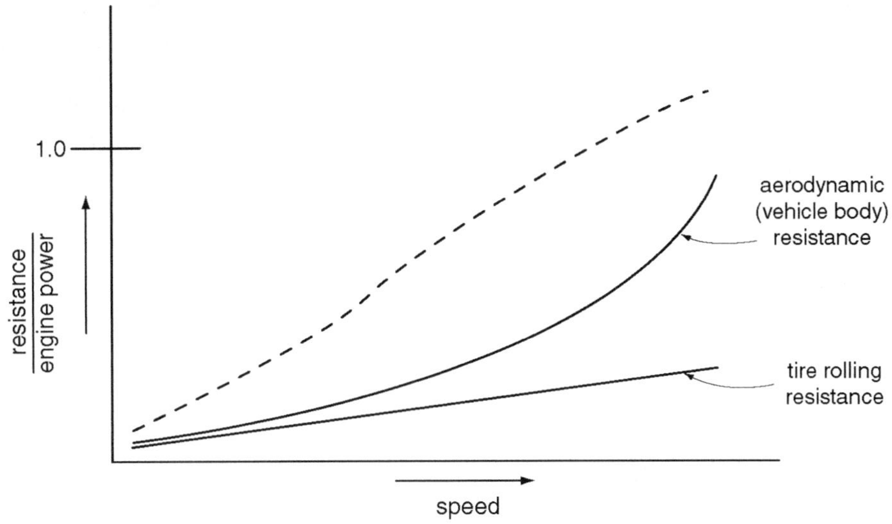

Fig. 5.5 Comparison of the tire rolling resistance and the vehicle body drag

5.3 Tire Contribution

The tire, being the contact device between the vehicle and the road, plays prominently in the study of vehicle performance. The tire supports the vehicle and enables the vehicle to respond to the actions—steering and braking—initiated by the driver.

In this section we will consider the effect of wheel rotation on air flow pattern. The wheels of a vehicle rotate in a direction opposing the airstream (Fig. 5.6).

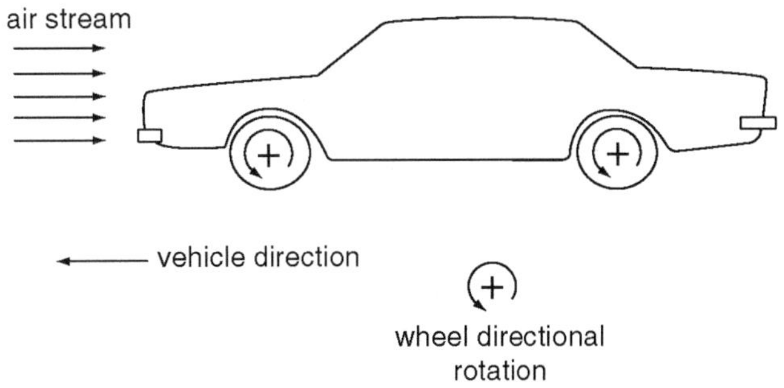

Fig. 5.6 Direction of tire rotation

The presence of wheels is responsible for a drag coefficient increase of up to 0.08 and a lift coefficient increase of 0.3 to 0.58 [5-5, 5-6]. Although the wheels do not produce the lift, they are responsible for the contribution to the lift. The rotating wheel itself induces a vortex flow. This effects the formation of a vortex, albeit weak, by the air mass in the proximity of the wheel. The airstream on the side of the car travels into this vortex, attempting to bend it as the resulting compound vortices fuse into the primary vortices created by the main body of the vehicle (Fig. 5.7(a)). The top of the wheel (at region T) moves into the airstream at twice the driving speed (Fig. 5.7(b)). It follows therefore that particles of air still travel in the direction of the car. This results in an increase in the aerodynamic drag, although the aerodynamic lift is reduced. The separation point ranges from a point S1 through T to a critical point S2. The higher the speed, the more forward is the separation point, up to the critical point, S2, beyond which a further increase in speed has no effect on the separation point.

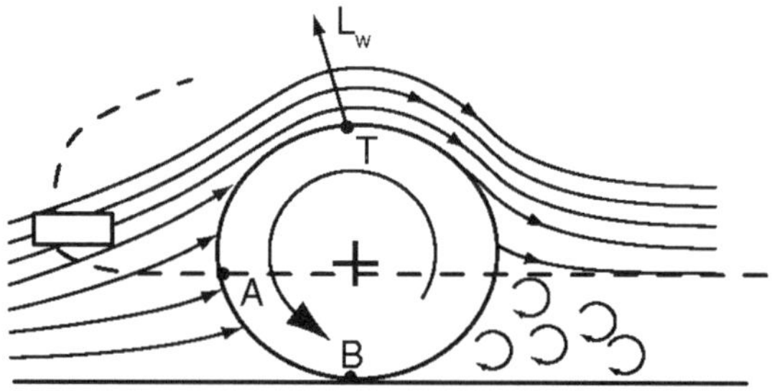

Fig. 5.7(a) Tire adds to vehicle lift and aerodynamic drag

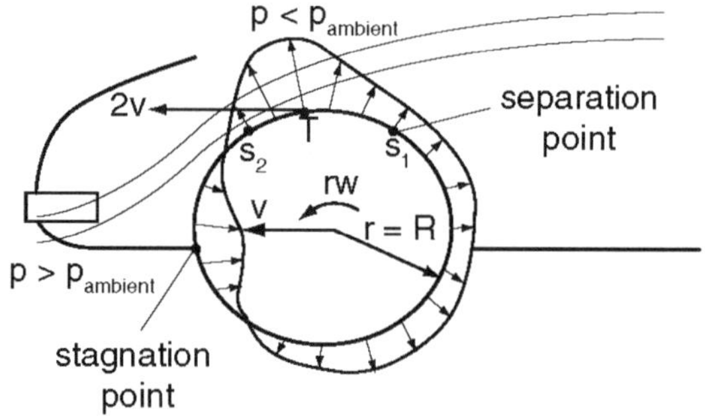

Fig. 5.7(b) Pressure distribution on road tire

The front wheels contribute the bulk (about 65%) of the total wheel drag in cars because the rear wheels are typically in the wake of the front wheels, only experiencing some of the vortices generated by the front wheel interaction with the car-body air flow, and the lower edge of the trailing stream flow from the front wheel (Fig. 5.8). This front wheel contribution is lower in trucks. In fact, the front wheels contribute less than 30% of the total wheel drag in trailers, and about 40% in buses.

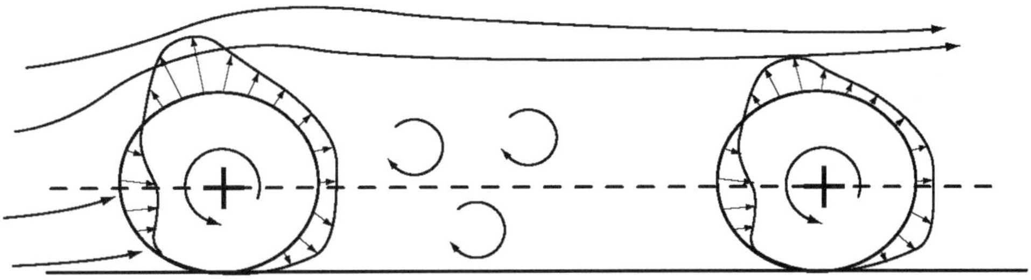

Fig. 5.8 Rear wheel in the wake of flow over front wheel

For a given ground clearance, beyond a critical value that is experimentally determined, drag and lift increase as the radius of the wheel-housing or its depth increases, the radius being the stronger influence on drag and on lift. Therefore, in recommending a minimum wheel-housing volume for a reduction in the lift and drag of a vehicle, emphasis should be placed on reducing the radius to achieve the reduction in volume [5-7]. A wider wheel contributes to an increase in lift and drag, but a larger radius wheel of the same boxed-in wheel volume contributes more to drag. This is consistent with the changes found in increasing wheel-well volume beyond the critical value. Up to that critical value, the drag reduces with increased radius and reduced clearance (Fig. 5.9). It has been determined that a shielded height-to-wheel diameter ratio of about 0.75 gives the minimum wheel drag [5-8].

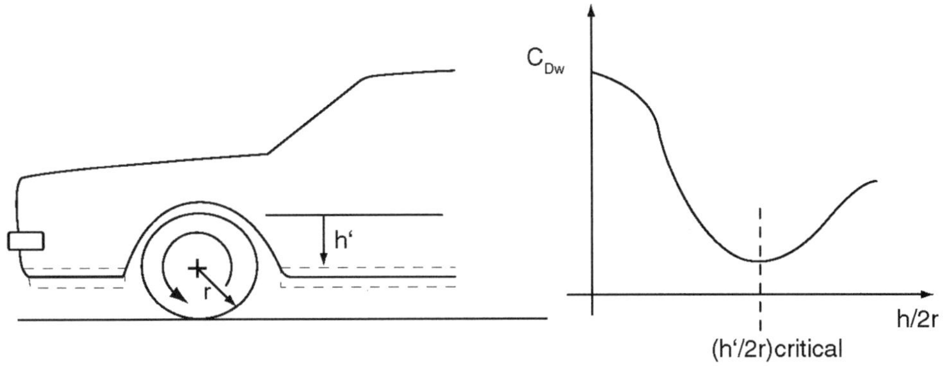

Fig. 5.9 Variation of wheel drag with wheel radius, *r*, and wheel cover, *h*

Although narrower tires give better aerodynamic characteristics (lower ΔCD), wider tires are better for performance because they have a higher ground contact area, which helps to improve road-vehicle interaction (Fig. 5.10). Even then, there are performance constraints, apart from the certain architectural constraints, to increasing the tire width. The width-to-diameter ratio should be kept below 1.0 for optimum benefit. Although the performance may improve up to about b/D = 1.5, the added weight requires more power, and the increase in lift and in the aerodynamic interference will negate the little gain in performance.

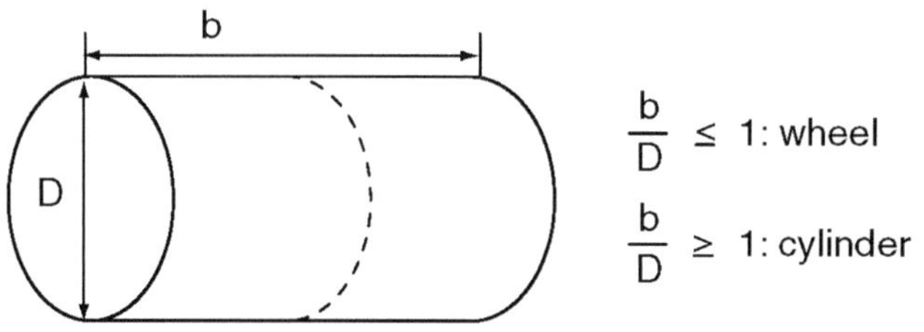

Fig. 5.10 Wide tires (up to b/D ≤ 1) improve performance

Some other aspects to consider in wheel design are the wheel cap hole and the air flap. The designer seeks to design for an increased area of the wheel cap hole. The larger the area of the wheel cap hole, the lower will be the braking temperature (Fig. 5.11). Air flaps serve to guide air smoothly over the wheels, thereby reducing the wheel aerodynamic drag. A well-designed air flap will slide the stagnation point from A to B in Fig. 5.7(a).

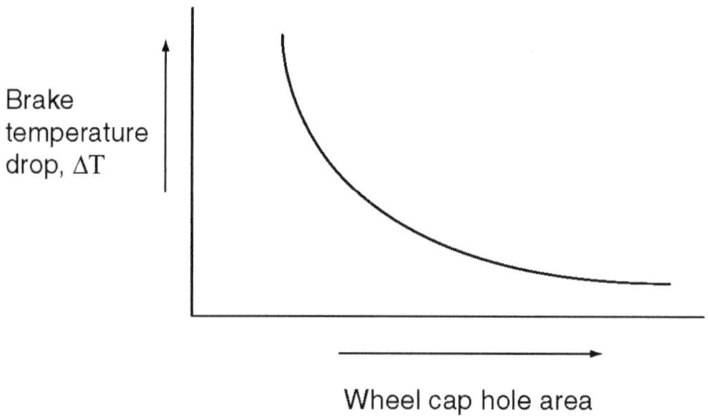

Fig. 5.11 Brake temperature drop versus wheel cap area

As noted in Section 3.1.12, the rotating wheel, which represents the true case of wheel operation, has a lower drag in wind tunnel testing than a stationary wheel. Testing with a stationary wheel could over predict the wheel contribution to vehicle drag by as much as a drag coefficient of 0.021.

The rolling resistance increases with speed for a given tire temperature (Fig. 5.12), up to about 70 mph (113 km/hr), and decreases with temperature for a given speed. The rolling resistance also decreases with the firmness of the road, with concrete roads having the lowest resistance (0.01) and unpaved roads having the highest (0.05).

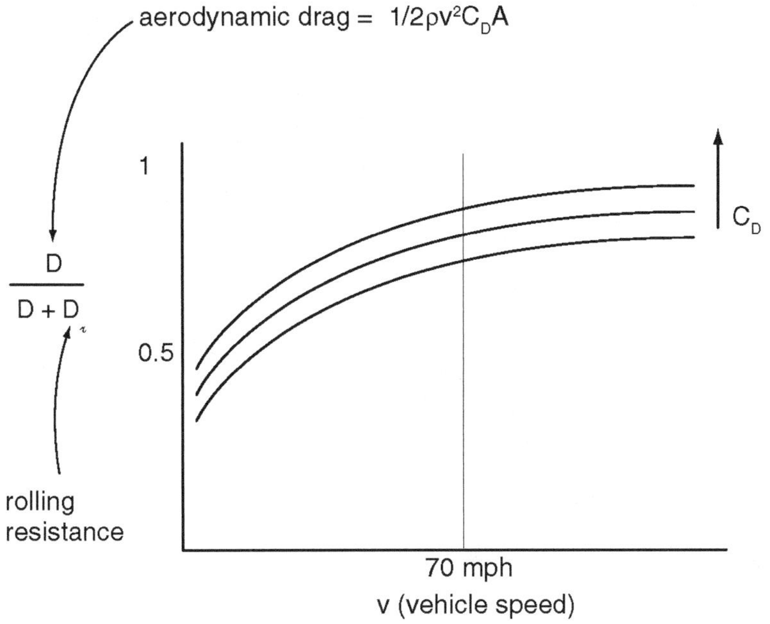

Fig. 5.12 Rolling resistance vs. vehicle speed

Tests on scale models must cover a wide speed range to obtain a reliable statement on the effect of tire rotation because (as we have seen) at low speeds the aerodynamic drag due to the rotation is significant (sometimes dominant), whereas at high speeds the vehicle body drag dominates and increases with speed while the tire drag remains flat.

As part of the effort to minimize wheel drag, car designs were at one time introduced in which the wheels were enclosed in the full body of the vehicle so that only the lower third or so of the wheel is exposed. Tests suggest that such a design actually adds to the vehicle drag [5-7, 5-9]. The enclosure side adds to the separation drag, and the enclosure bottom adds to the lift due to interference.

5.4 Sideslip and Turning

In a turn, a vehicle develops a slip angle, at the tire, relative to its heading. This sideslip of the tire introduces a proportional side force. The less the sideslip angle, the less is the side force, and the closer is the direction of travel aligned with the direction to which the tire points. A true turn occurs when all the wheels of a vehicle maintain a common sideslip angle, thus maintaining a common tangential velocity to the circular path of the turn, hence a four-wheel drive. Oversteering (also referred to as tail-happiness), a condition in which the slip angle of the rear tires is greater than that of the front tires, tends to be more likely in rear-wheel drive vehicles and understeering in front-wheel drives (Fig. 5.13).

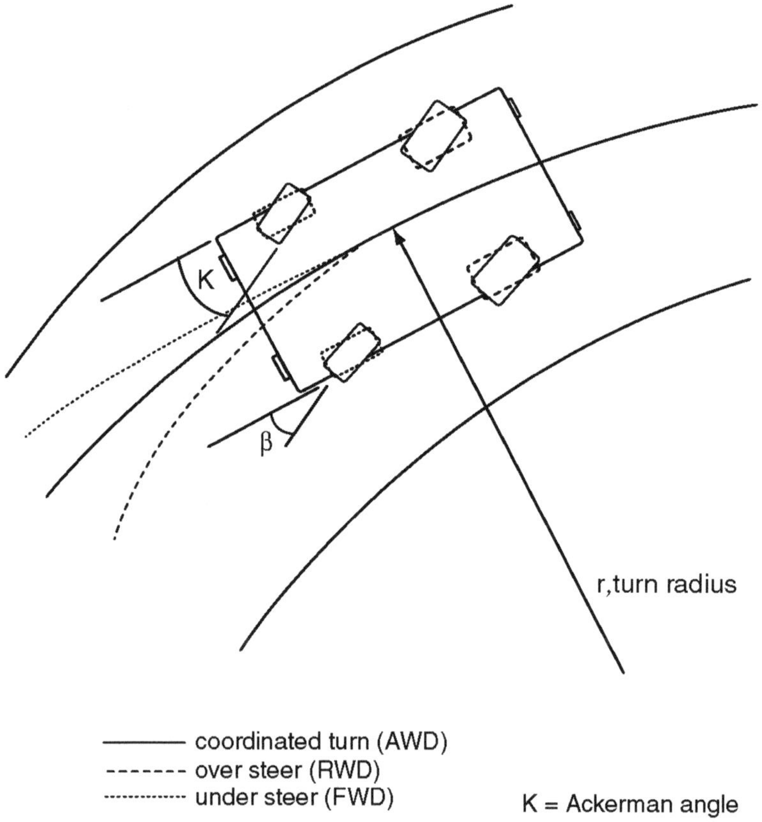

	coordinated turn	over steer	under steer
wheel combination front and back	1-1	1-2	3-1
comparison with ackerman angle	$K=K_C$	$K>K_C$	$K<K_C$
comparison with sideslip angle	$\beta_{front}=\beta_{back}$	$\beta_{front}>\beta_{back}$	$\beta_{front}<\beta_{back}$

Fig. 5.13 Cornering

5.5 Trailing Vortices

The aerodynamic lift is proportional to the speed of circulation. Figure 5.14 shows the vortex flow trailing a car.

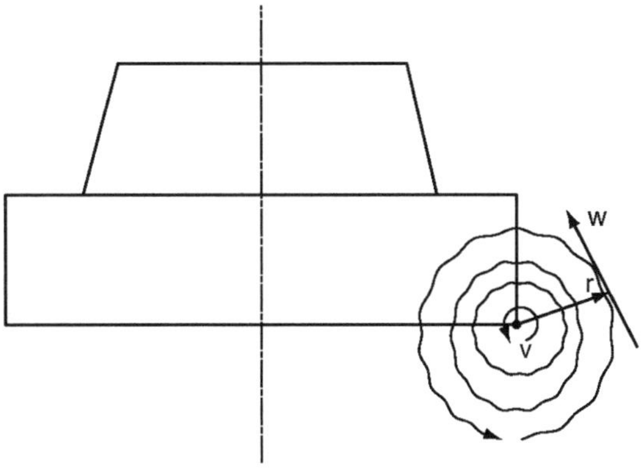

Fig. 5.14 A trailing vortex

$$\Gamma = 2\pi r w$$

where

Γ is the circulation;

r is the radius of the circular streamline within the vortex; and

w is the tangential velocity produced by a rotating motion.

From conservation principle, the main vortex has the same strength as the trailing vortex before dissipation. Therefore the trailing vortex can be analyzed in terms of the well-defined main vortex. The lift due to the main vortex is a function of Γ, and is given by

$$L = \rho \mathrm{v} \Gamma d = \rho \mathrm{v} 2\pi r w d$$

where d is the width of the car rear.

Recall the general aerodynamic expression, $\frac{1}{2}\rho \mathrm{v}^2 C_L A$, for lift.

Equating both lift expressions, we get

$$C_L = \frac{4\pi r w d}{\mathrm{v} A}$$

At some point in the trail, the width, d, of the car is approximately $2r$. Substituting for the lifting area ($A = 2rd$) the lift coefficient becomes

$$C_L = \frac{2\pi w}{\mathrm{v}}$$

This approach is sometimes used to estimate the lift coefficient. One can take advantage of well-developed vortices to delay flow separation on the rear surfaces of a car. Such a beneficial trailing vortex can be obtained by styling the car such that the rear surface has a gentle, negative slope leading to the bobtail. The aim of the gentle slope design is to avoid sudden flow separation, and the goal of bobtailing is to keep the suction in the wake small. As shown in Fig. 5.14, a wake is formed at the rear and surrounds the surface situated downstream of the separation boundary layer.

Consider for a moment that the entire car is an airfoil represented by Fig. 5.15. Cutting off the rear end does not change the resultant aerodynamic force on the vehicle. It only replaces natural separation by forced separation. The local pressure within the wake is practically uniform and below atmospheric, hence there is suction into the wake.

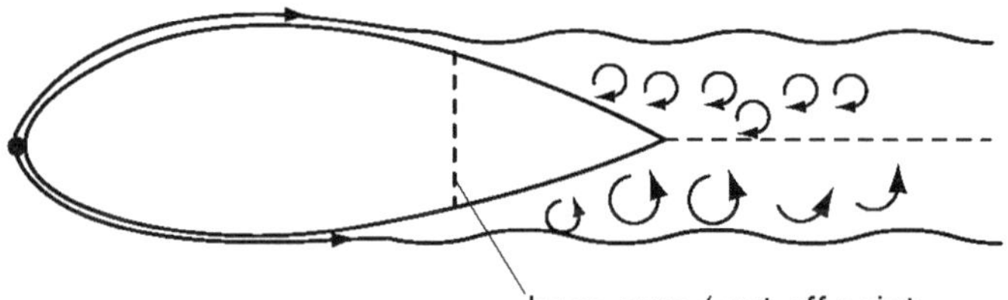

Fig. 5.15 Bobtailing

5.6 Control Surfaces

Control surfaces are discussed further chapter 10. The topic is briefly introduced in this section. If properly placed, control surfaces (spoilers, winglets, end plates) serve to reduce vehicle aerodynamic drag. Drag reduction of between 0.015 and 0.023 can be obtained from adding spoilers to the basic car body [5-9].

The induced drag from the lift-generating devices usually outweighs the lift benefits. As a result, the performance, a measure of lift-to-drag ratio, depreciates. In the absence of lift, (the normal force), the driving force on a vehicle is constant. But when aerodynamic lift is present, the normal force changes with speed in the form $F = kv^2$, and the driving force is increased for downforce (negative lift) and decreased for lift. Whether there is an improved performance depends on the difference between the aerodynamically influenced downforce (or lift) and the induced drag of the vehicle. A relatively higher driving force translates into an improvement in performance.

5.7 Car Underbody Flow

The closer the vehicle is to the ground, the greater is the significance of the underbody flow. Because of the benefit of downforce, it would seem desirable to design a car to be as close to the ground as possible. However, because of the viscosity inherent in the flow, the benefit of the downforce would be negated if the clearance is below critical ground clearance. The underbody flow is actually three-dimensional as it spreads out laterally along its longitudinal main flow. The flow near the ground is greatly influenced by viscosity. The closer the car is to the ground, the more trapped is the air under the vehicle, resulting in a slower air flow at a higher pressure than the surrounding air. This underbody higher pressure air does two things that are detrimental to drag reduction on the car. First, it flows to the sides of the car, creating a vertical component that obstructs the otherwise longitudinal flow on the car. The effect is like the tripping of flow on a surface to arrive at premature turbulence. Second, the high-pressure flow swirls around the lower sides of the car, creating additional lift, rather than the desirable downforce, due to the circulation. In aiming to keep the car close to the ground, it should not be so low as to trap the air between the underbody and the ground. Rather, it should be just high enough to allow for an accelerated flow, which causes a reduction in aerodynamic lift—an increase in downforce.

The theoretical critical clearance is the height at the mixing length. The mixing length is the distance from the vehicle leading edge to the merge of the respective boundary layers on the car underbody and on the ground (Fig. 5.16). The flow due to this mix adds to the turbulence, resulting in lift under the vehicle. The further the mixing length is toward the rear of the vehicle, the lower is the critical clearance. When the mixing length is extended, the rate at which the boundary layer grows reduces, hence the drag incurred due to ground effect is less. A careful styling of the car fascia can contribute substantially to a smooth underbody flow since the underbody is, at least aerodynamically, a near-seamless continuation of the vehicle's front face.

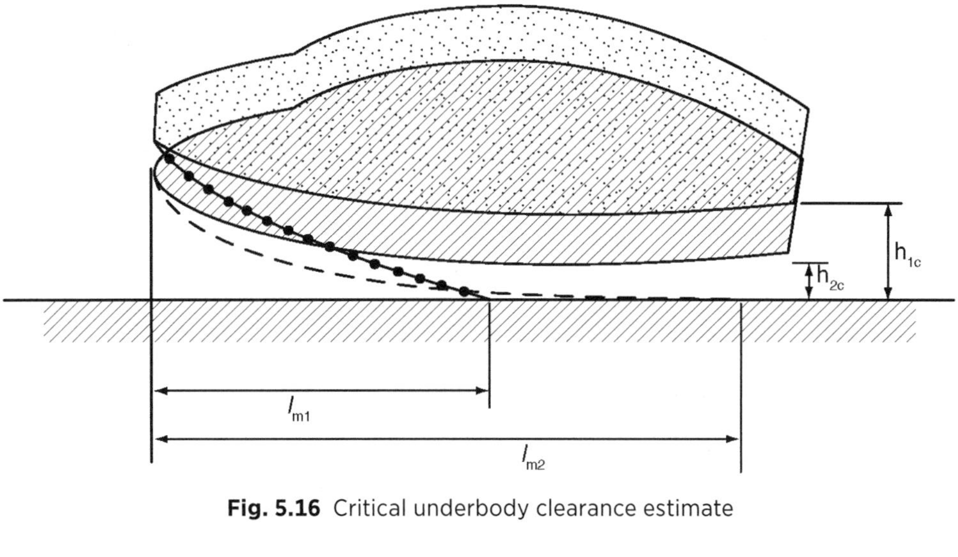

Fig. 5.16 Critical underbody clearance estimate

In reality, the critical clearance will need to be higher than the smooth-underbody estimate because of the exposure to the airflow of components—exhaust system, shock absorber, gas tank, and numerous cables—attached to the underbody.

5.8 Cross Wind

The behavior of road vehicles in cross wind is a rather complex phenomenon involving vehicle aerodynamic build, vehicle suspension system, the braking system, the steering wheel, and even (perhaps particularly) the driver. In spite of all the intelligent maneuvering systems built into the modern car, an incompetent driver makes the vehicle as vulnerable as though there were no in-built precautions—this is a lesson that airplane pilots have clearly learned.

Yaw is the angle that the relative wind makes with the vehicle's longitudinal direction. Experiments in the study of the effects of cross wind have shown that the data obtained from steady-state dynamic tests are comparable to those obtained from static tests [5-10]. The rush of air around the corner of the vehicle causes a lateral low-pressure zone that is primarily responsible for the yaw at the vehicle front wheels. Yaw contributes to a significant increase in the drag at the wheels. The coefficient of yaw (side force) is given by

$$C_Y = \frac{Y}{\frac{1}{2}\rho v^2 A}$$

where

Y is the side force and
A is the frontal projected area.

The projected frontal area is still used for consistency in nondimensionalization, even though the projected area exposed to the side force is the vehicle's side area.

The coefficient of yawing moment is given by

$$C_N = \frac{N}{\frac{1}{2}\rho v^2 A l}$$

where

N is the yawing moment and
l is the longitudinal distance between the wheels.

Aerodynamic yawing moments are typically measured about an axis located at the wheelbase center. The direction of yaw response to side force is determined by the moment about the center of gravity, however, not about the suspension geometry.

5.9 Truck-Car Passing Interaction

Consider a car speeding ahead to overtake a truck (Fig. 5.17). As the car approaches the truck in a relatively calm wind condition, the car is pushed away laterally from the truck, and the driver has to steer toward the truck to counteract this turning effect. For the duration of the car driving beside the truck, there is no noticeable push or pull of air until the car is about to pull ahead of the truck. Then the car driver feels a pull toward the truck—a scary experience, since the wheels of the truck are standing higher than the eye level of the car driver. The car driver must steer away from the truck when pulling out of the truck environment.

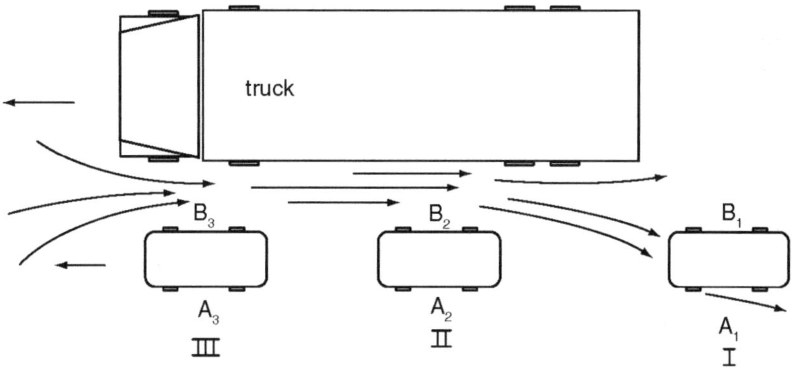

Fig. 5.17 Car overtaking truck

Interestingly, a few articles have been published that submit the opposite of the preceding description, based on wind tunnel tests. However, repeated tests on the highway have yielded the same experience described. An explanation of the interaction between the car and the truck is presented next.

Stage I: The airstream hits the car, which at this time is in its lane, longitudinally behind the truck. There is only one wall—the truck. The air flow off the truck hits the car head-on, and a portion of it flows past side B1 of the car. The only surface—at least one of minimum resistance—available to the car is side A1. Therefore, the car naturally tends to yaw toward A1. The driver compensates by steering the car toward B1 to maintain a longitudinal motion.

Stage II: The airstream between the car and the truck is only in the longitudinal direction. The tube of flowing air is constricted here to flowing faster at B2 than at A2. One would therefore expect suction toward the truck, yet this is not the case, perhaps because of the lateral component of the flow from the wheels and the boundary layer build-up between the two vehicles.

Stage III: When the car is about to pull from the influence of the truck, there is a strong feeling of being sucked in. This is perhaps because the flow between the truck and the car is constricted the most at this point, in effect hemming in the air between the two walls, and preventing the scattering of the incoming air stream. The speed of the constricted air is greatest at this point (B3), and, correspondingly, the pressure is the least ($p_{A_3} >> p_{B_3}$). Hence, the pronounced suction toward the truck. The strength of the suction depends on the relative speed between the car and the truck. Also, the compressed air just ahead of the truck creates a wall in front of the car. The result of the two effects is a pull of the car toward the truck (Fig. 5.18). This phenomenon has been observed to be consistent for all three cases of tractor-trailer combination, tractor-trailer bed (without trailer), and tractor alone. This leads to conclude that the car-truck interaction as the car leaves the side area of influence of the truck (Fig. 5.19) is predominantly a function of the truck front design.

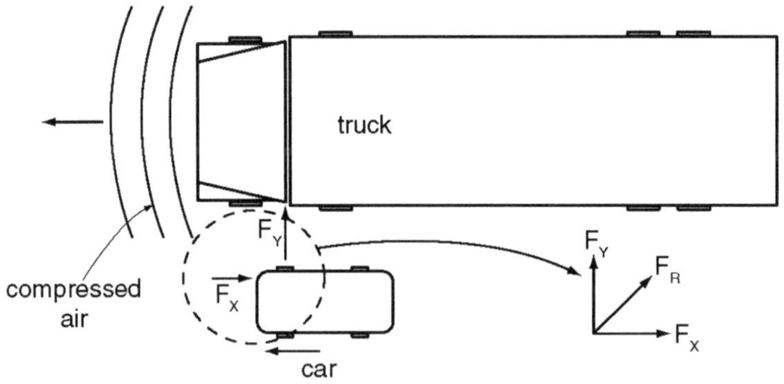

Fig. 5.18 Car pulling out of truck influence

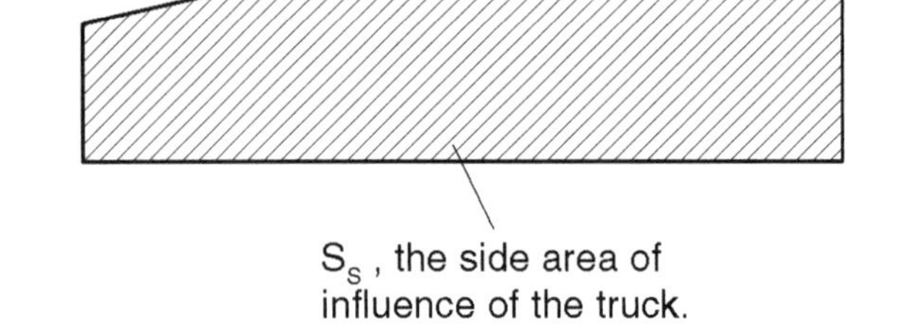

Fig. 5.19 Truck side section influencing passing car

The truck-car interaction described is for calm wind conditions on a straight stretch of the highway. The effects of bends, undulating roads, cross wind, head wind, and tail wind change the car-truck interaction in a way that is yet to be fully understood, especially when a combination of those effects occurs simultaneously.

5.10 Effect of the Environment on Aerodynamic Drag

Just as with density, aerodynamic drag reduces at higher temperatures and at higher altitudes. The similarity to density variation is to be expected since the aerodynamic drag is a function of the density—$D = \frac{1}{2}\rho v^2 C_D A$. An increase in ambient temperature from 32 °F to 100 °F (0 °C to 38 °C) will cause a 14% reduction in the aerodynamic drag because of the corresponding reduction in density. Moving to altitude 4000 ft (1219 m) from sea level will cause a decrease of 17% in the aerodynamic drag. An ideal region for aerodynamic drag reduction would therefore appear to be a hot highland. Unfortunately, the higher one goes, the cooler it becomes. Besides, as the temperature of the environment increases, there is fuel loss due to evaporation. Also, fuel becomes more expensive, albeit not by design, since fuel is sold by the volume and not by the mass.

For gas engines at wide-open-throttle (WOT), the engine power at operating condition is given in terms of the standard atmospheric condition [5-11] as

$$p = \frac{p_0(p_{BAP} - p_{vapor})}{p_{SDABP}}\sqrt{\frac{T_0}{T}}$$

where

p_0 is the reference atmospheric pressure;

p_{BAP} is the barometric pressure at engine performance;

p_{SDABP} is the standard dry air barometric pressure (101.32 kPa);

T_0 is the standard (absolute) reference temperature; and

T is the ambient (absolute) temperature at engine performance.

5.11 Effect of Aerodynamic Drag on Fuel Consumption

The tractive force supplied by the engine is employed in overcoming the aerodynamic drag, inclusively. Therefore the drag affects the power required. The latter is expressible in terms of the fuel consumption.

The minimum drag is obtained when there is a balance struck between the form drag and the skin friction drag. The skin friction drag increases more rapidly with length than the form drag, which is a normal pressure distribution. Therefore, drag reduction and the consequential reduction in fuel consumption can be achieved by bob-tailing.

Summary

Vehicle performance is indicated by the functional characteristics of the vehicle. Those characteristics include the ability of the vehicle to accelerate and decelerate quickly, its torque for climb or for load pull, and its ride.

The power required for the operation of a vehicle is made up of mechanical power and aerodynamic power. Vehicle speed is a major determinant of the required aerodynamic power—the mechanical power varies with the vehicle speed but the aerodynamic power is proportional to the cube of the vehicle speed.

The tire, when rotating, induces a vortex flow that contributes to lift in ground vehicles. Tire design or selection is therefore of aerodynamic significance.

Key Terms

Aerodynamic power: The component of the engine power required to overcome aerodynamic drag.

Coefficient of road adhesion: The ratio of the vehicle tractive (braking) force to its normal load; it is strongly dependent on the road-tire condition

Control surfaces: Devices attached to the vehicle main body to reduce drag and to improve vehicle performance; spoilers are a common control surface device.

Cross wind: Wind that blows such that it has a lateral contribution to the (longitudinal) direction of travel of the road vehicle; cross winds contribute to vehicle instability. Although a cross wind can be tail cross wind or head cross wind, it is generally assumed to be a head cross wind except otherwise stated.

Head wind: The wind that blows head-on, directly opposite the sense of travel of a vehicle as the vehicle travels in its path.

Mechanical power: The component of engine power required to overcome traction; total power required is the sum of the mechanical power and aerodynamic power.

Performance: A description of the dynamic behavior of a vehicle, including its ability to accelerate and decelerate quickly, its torque for climb or for load pull, and its ride.

Stability: The ability of a vehicle to maintain or return to its equilibrium state; stability in ground vehicles refers, in general, to dynamic stability.

Tail wind: The wind that blows from the rear, in the sense of travel as the vehicle as the vehicle; tail wind is aerodynamically favorable to vehicles because it causes a reduction in the aerodynamic drag.

Review Questions

1. Define vehicle performance.
2. What is traction in ground vehicle? What load component of the vehicle is the most important when considering vehicle traction?
3. Describe the traction-limited phase in vehicle performance.

Application Questions

1. You are traveling northbound on a five-lane highway in a mid-size sedan. You are on the fourth lane. A slow truck is in the third lane. You intend to stay in your lane as you approach with the intent to pass the truck. There is a cross wind at 30° southeast wind confronting you and your parallel friend, the truck driver. Describe the forces that are at play (a) as you enter the truck environment, (b) while in the truck environment, and (c) as you exit the truck environment. How do you expect these forces to affect the stability and performance of your car? What corrective steering measures should you take in each of three phases?
2. Extend Eq. 5.17 to determine the tractive force in a tandem drive tractor trailer. (Hint: Treat the last trailer as a dead weight and add its drag contribution to R_d when treating the first trailer.)

Exercises

1. Using Eqs. 5.10a and 5.10b, if the center of gravity is midway between the front and the rear tires, compare, for a mild incline, the maximum tractive forces for a front-wheel drive and a rear-wheel drive. (Coefficient of road adhesion is 0.8.) What is the vehicle normal load? Would you design this vehicle as a rear-wheel drive or as a front-wheel drive? Explain.
2. Can the coefficient of road adhesion be grater than 1? Explain.

3. Using Eq. 5.21, compare the power required to overcome aerodynamic drag to that required to overcome tractive force. Plot the comparison as a function of velocity to 150 mph.

Project

Set up two tractor-trailer arrangements, one a full tractor-trailer, the other a semi-tractor trailer. Be sure that both have the same cab design and the same length (from the front of the tractor to the rear of the trailer). With both trucks empty, and for the same road conditions, compare the aerodynamic drag of both designs as a function of speed.

Compare the stability of a sports car with that of a sedan in the following scenarios:

a. Overtaking a tractor trailer
b. Overtaking a low-bed (no carriage) tractor trailer

References

1. Hawks, R. J., and Sayre, C. L., "Aerodynamics and Automobile Performance," *Advances in Road Vehicle Aerodynamics*, Stephens, H. S., ed., BHRA Fluid Engineering, Bedford, England, 1973.
2. Gillespie, T. D., *Fundamentals of Vehicle Dynamics*, SAE International, 1992.
3. Katz, J. K., *Race Car Aerodynamics: Designing for Speed*, Robert Bentley, Inc., 1995.
4. Wong, J. Y., *Theory of Ground Vehicles*, 3rd ed., John Wiley & Sons, Inc.,2001.
5. Carr, G. W., "Potential for Aerodynamic Drag Reduction in Car Design," *International Journal of Vehicle Design, Technological Advances in Vehicle Design Series, SP3, Impact of Aerodynamics on Vehicle Design*, 1983, pp. 51–52.
6. Mercker, E., and Knabe, H. W., "Ground Simulation with Moving Belt and Tangential Blowing for Full-Scale Automotive Testing in a Wind Tunnel," SAE Paper no. 890367, 1989.
7. Cogotti, A., "Aerodynamic Characteristics of Car Wheels," *International Journal of Vehicle Design, Technological Advances in Vehicle Design Series, SP3, Impact of Aerodynamics on Vehicle Design: Technological Advances in Vehicle Design, SP3*, ed. 1983, pp. 173–196.
8. Scibor-Rylski, A. J., *Road Vehicle Aerodynamics*, John Wiley & Sons, 1984.

9. Mercker, E., Bauer, H., Bernaburg, H., and Emmelmann, H. J., "On the Aerodynamic Interference Due to the Rolling Wheels of Passenger Cars," SAE 910311, Vehicle Aerodynamics, Recent Progress, SAE S-P 855, 1991.

10. Minato, K., and Ryu, H., "Aerodynamics of Vehicles in Tunnels—Flow Visualization Using Laser Light Sheet Method and Its Digital Image Processing," SAE Paper 910314, 1991, pp. 123–131.

11. Tabouk, J. J., "Mechanics of Vehicles," *Machine Design*, May 30–Dec.27, 1957.

Chapter 6

Vehicle Sectional Design

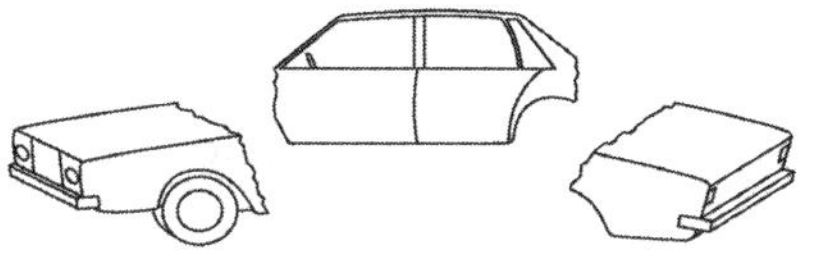

The growth of viscous boundary layer on the surface of a vehicle depends on such factors as the local free stream velocity induced by the shape and the inclination of the local surface to the free stream [6-1, 6-2, 6-3]. The surfaces of a car are typically of a 3-D curvature, with most of the variation in surface air flow taking place along the longitudinal axis (Fig. 6.1). The body shape of a car can be divided into three parts—front, mid, and rear—by a plane normal to the general direction of motion. The maximum cross section is in the first part. Air particles accelerate and the dominant flow stays attached to the surface. It

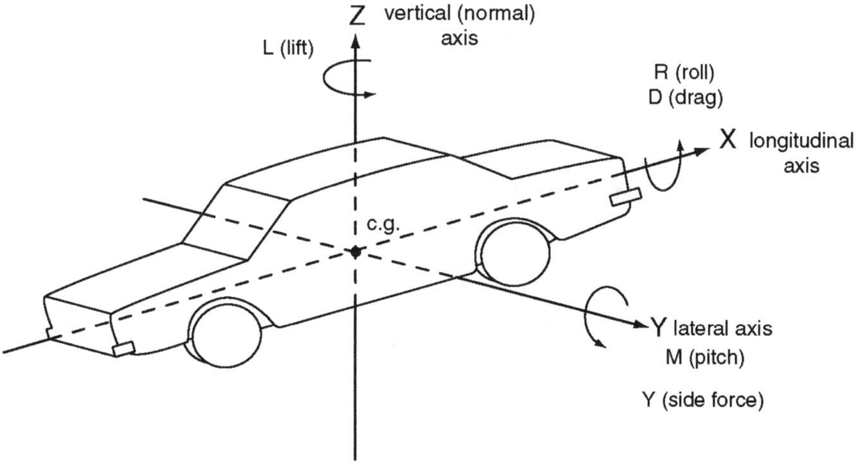

Fig. 6.1 Surfaces of a car: 3-D

would take artificial means—such as a sharp turn or cross wind—to cause flow separation on this portion of the vehicle surface. At the front of the vehicle the air flow is upward, following the surface, into the free stream. The accelerating air travels past the hood and the windshield and continues over the front portion of the passenger compartment. Some separation—mini separation—occurs at the hood-windshield interface, at the windshield-roof top interface, and around the pillars (Figs. 6.2, 6.3). But the flow soon reattaches and the local pressure continues to decrease.

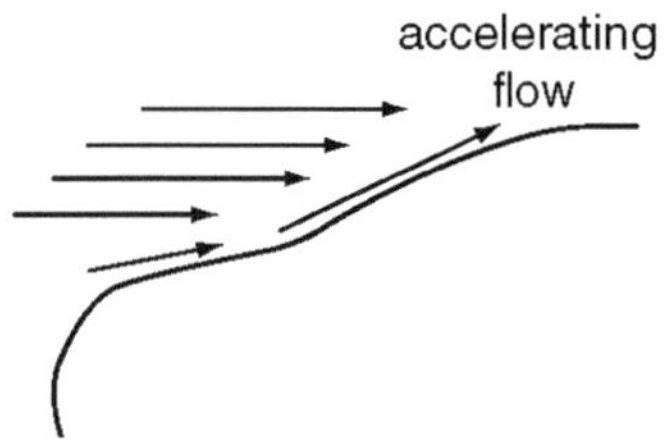

Fig. 6.2(a) Car front centerline

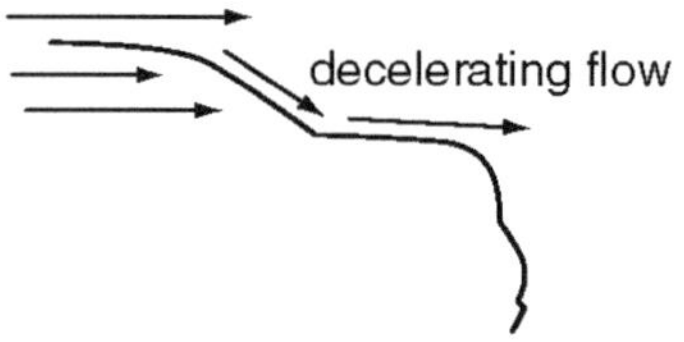

Fig. 6.2(b) Car rear centerline

The flow continues over the top and around the sides toward the rear of the vehicle. It is at the midsection (also referred to as the cabin) that the flow starts to decelerate. The cross section gradually reduces toward the back, slowing down the flow and creating a condition that facilitates flow separation. Flow separation is inevitable, but it can be controlled. It is better to induce flow separation where it is least obstructive.

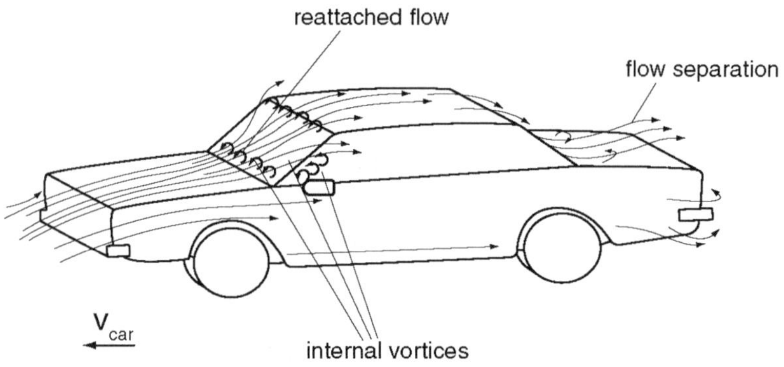

Fig. 6.3 Leading section flow dynamics

The rear surface is a continuation of the new flow properties—decreasing speed, increasing pressure—acquired at the midsection. At the rear, the surface curves down farther, away from the free stream; the air decelerates, and the local pressure increases. Air flows against increasing pressure causing an adverse (or positive) pressure gradient and a potential for boundary layer separation (Figs. 6.2(b), 6.3). The drag coefficient due to the body shape, the profile drag, can be reduced from the typical 0.28 to about 0.1—a 64% drag reduction—by a suitable combination of the front, mid-, and rear design shapes [6-4].

6.1 Car Front and Mid Sections

The flow is unlikely to separate on the front of the car body except at the meeting point of the hood and the windscreen. Flow is fast over the hood, but the windscreen slows it down. The steeper the angle the windscreen makes with the hood, the slower is the flow. This slowing of the air stream causes it to separate in the front of the windscreen, resulting in separation bubbles at the interface of the windscreen and the hood. The flow reattaches at the upper part of the windscreen and continues its streamline pattern to the roof of the car. The mini-separation that occurs at the interface of the hood and the windscreen is a 3-D phenomenon, with vortices extending sideways and carried along some distance on the sides of the vehicle (Fig. 6.3). The flow is affected by three main factors: the angle between the hood and the windscreen, the curvature of the hood, and the curvature of the windscreen.

6.1.1 Effect of γ on the Flow

For simplified analysis, consider the hood and the windscreen to be flat surfaces, with angle γ between the windscreen and the projection of the hood (Fig. 6.4). For $\gamma \leq 60°$, the separation point on the hood and the reattachment point on the windscreen are respectively given by

$$\frac{x_s}{l_h} = 0.585 + \left[0.172 - \left(\frac{\gamma}{150}\right)^2\right]^{0.5} \tag{6.1}$$

and

$$\frac{x_R}{l_w} = 0.415 - \left[0.172 - \left(\frac{\gamma}{150}\right)^2\right]^{0.5} \tag{6.2}$$

where l_h and l_w are the respective lengths of the hood and the windshield. The separation and reattachment points are determined by the combination adjustment of windscreen and the hood to yield the included angle, γ.

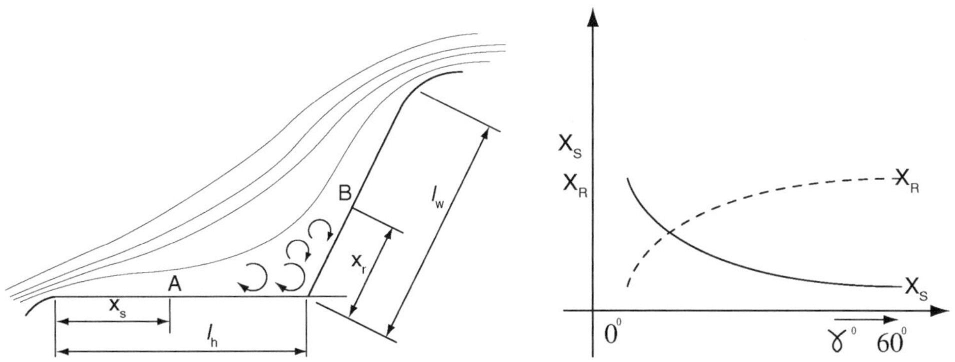

Fig. 6.4 Separation point and reattachment point

6.1.2 Effect of Hood and Windscreen Curvature

Increased curvature of the hood tends to move the separation point downstream. Likewise, increased curvature of the windscreen creates the beneficial tendency of moving the reattachment point to the lower end of the windscreen (Fig. 6.5). Increased camber of the hood or of the windscreen will cause a delay in separation and in reattachment (Fig. 6.6). Side flow restrictions such as headlamps and side trimmings reduce the separation distance x_s because they contribute to early separation.

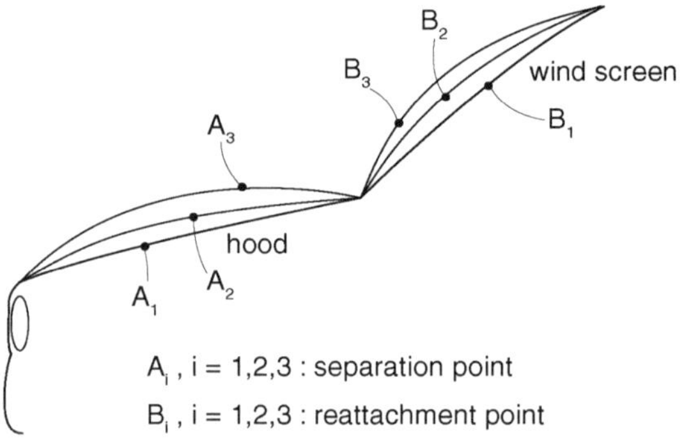

Fig. 6.5 Hood and windscreen curvature

The ideal combination would be to increase the hood camber and reduce windscreen camber to minimal shape for reattachment. Modify sides of windscreen camber for added curvature for the side flow. Because of the typical positioning of air inlet grills for ventilation at the base of the windscreen, the task of achieving the maximum x_s and minimum x_R becomes more intricate.

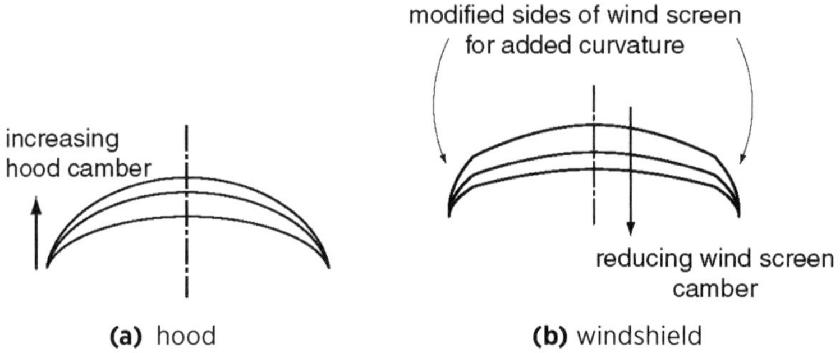

Fig. 6.6 Camber

6.1.3 Nose and Fascia Design

An approximate expression for the frontal area, A_f, of a passenger car in the small and medium car range is

$$A_f = 1.6 + 0.00056(m_v - 765)m^2 \tag{6.3}$$

where m_v is the vehicle mass in kilograms [6-5].

The goal is to make the frontal area wide and smoothly blending to the hood and to the vehicle underbody. Although Fig. 6.7(a) has a good frontal area width-to-height ratio, Fig. 6.7(b) meets this requirement better with the smooth blend of the frontal area up to the hood and down to the underbody, a design that allows for a curved radiator. A curved radiator has a wider surface area than a straight one of the same frontal area; this wider opening helps to reduce the interior pressure on the heating and cooling system.

The downforce is enhanced (or at least the lift tendency is reduced) by lowering the vehicle nose to just above a nose height that corresponds to the critical underbody clearance. A low vehicle nose usually requires lowering the radiator. A cross-flow radiator is a better fit than a down-flow one for the architectural and aerodynamic constraints of an optimum car leading edge because it has a relatively high width-to-height ratio. It does have the disadvantage of poor de-aeration compared with the down-flow radiator. Nonetheless, it is a preferred design for aerodynamic benefit. The de-aeration concern can be addressed in the design of the cooling system. The down-flow radiator would be a reasonable choice for a truck, because even with a lowered nose truck, the ground clearance is sufficient to accommodate a down-flow radiator within the aerodynamic profile developed from the leading edge.

The third bumper in Fig. 6.7(c) is selected because it is the most aerodynamic. The lower the front bumper, the less is the drag. So, for the front bumper, a combination of aerodynamic component shape and low placement within the vehicle front aerodynamic profile is needed to reduce drag. The C_D can be reduced by as much as 0.025 by a proper design and placement of the front bumper.

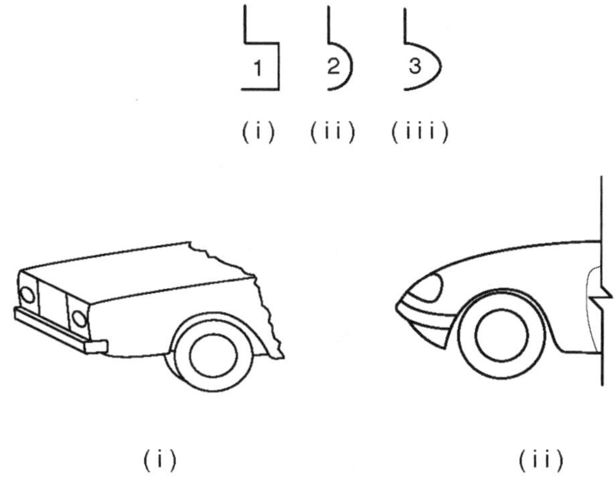

Fig. 6.7(a) Front section

6.1.4 Roof and Mid-section Pillars

As shown in Fig. 6.3, separation occurs at the interface of the windshield and the hood, at the blend of the windshield to the roof, and around the A-pillar. The B, C, and D posts, the door handle, the rear-view mirror, and the roof are among the components that make up the car mid-section or cabin. The A-post marks the lower boundary of the modified side of the windscreen in Fig. 6.6(b). The B, C, and D posts should be designed to be flush with the side glass and have an aerodynamic profile as shown in the cross-sectional view of Fig. 6.7(b), which is favored over Fig. 6.7(a). The door handle does not need to be flush with the car body to be aerodynamically beneficial; it only needs to blend with the car body in the same way that the posts blend with the side glass.

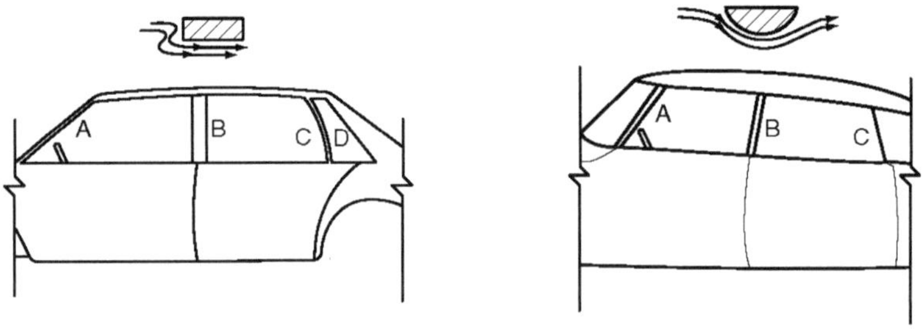

(i) (i,i)

Fig. 6.7(b) Midsection

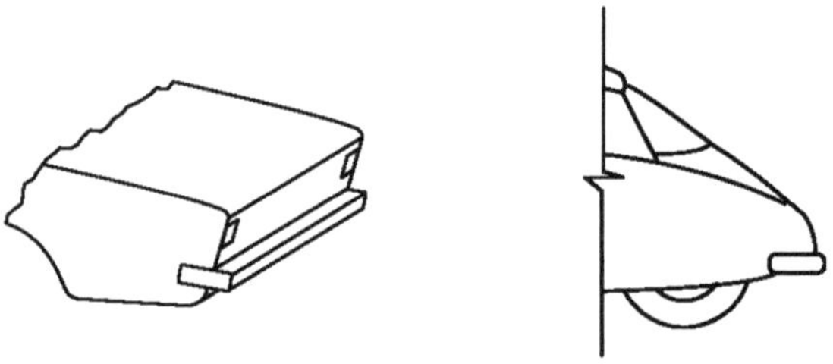

(i) (ii)

Fig. 6.7(c) Rear section

For the roof, a slight arching of the roof for a camber-to-chord length of up to 5% would help to delay flow separation (Fig. 6.8).

$$t^{*} = 0.05c$$

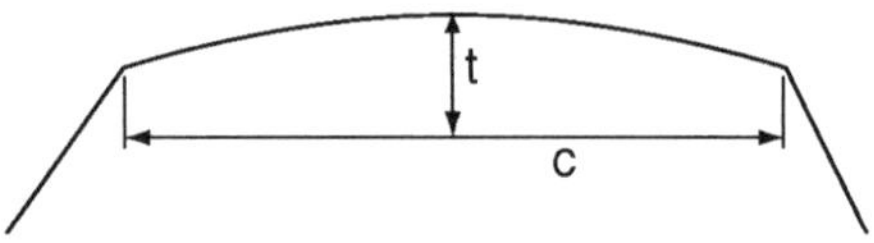

Fig. 6.8 Roof camber

6.1.5 Side-View Mirror

To determine the contribution of the side-view mirror to the vehicle drag, the total vehicle drag, D, is first obtained; then the drag, D', without the mirror is measured. Finally, the drag, D_m, obtained with just the mirror in a wind tunnel is recorded. It will be found that the total drag is greater than the sum of the drag without the mirror and the drag of the mirror, that is

$$D > D' + D_m \tag{6.4}$$

The drag D_m is defined by

$$D_m = \frac{1}{2}\rho v_L^2 C_{Dm} S_m$$

where

v_L is the wind tunnel (local) speed of test for the mirror and
S_m is the mirror frontal area.

The vehicle total drag is

$$D = \frac{1}{2}\rho v^2 C_D S.$$

Nondimensionalizing Eq. 6.4 with the dynamic head, $\frac{1}{2}\rho v^2$, and rearranging, we obtain

$$\Delta C_D = C_{Dm}\frac{v_L^2}{v^2}\frac{S_m}{S} + \Delta\Delta C_D \tag{6.5}$$

where

$\Delta C_D = (C_D - C_{D'})$, the drag contribution term due to the existence of the mirror; $C_{Dm}\frac{v_L^2}{v^2}\frac{S_m}{S}$ is the primary drag coefficient of the mirror; and $\Delta\Delta C_D$ is the additional drag term due to the secondary interference effect of the mirror.

This approach holds for all protrusions outside a vehicle's aerodynamic profile. Therefore, the same painstaking approach is needed to determine the drag contribution due to the door handles, headlamps, side lamps, hood ornaments, and antenna.

The further effect of secondary interference is more pronounced as the more conspicuously the protrusion deviates from the aerodynamic profile of the vehicle, but it reduces as the component is separated and distanced from the vehicle bulk profile, because it is then removed from the main flow regime.

6.2 Car Rear Section

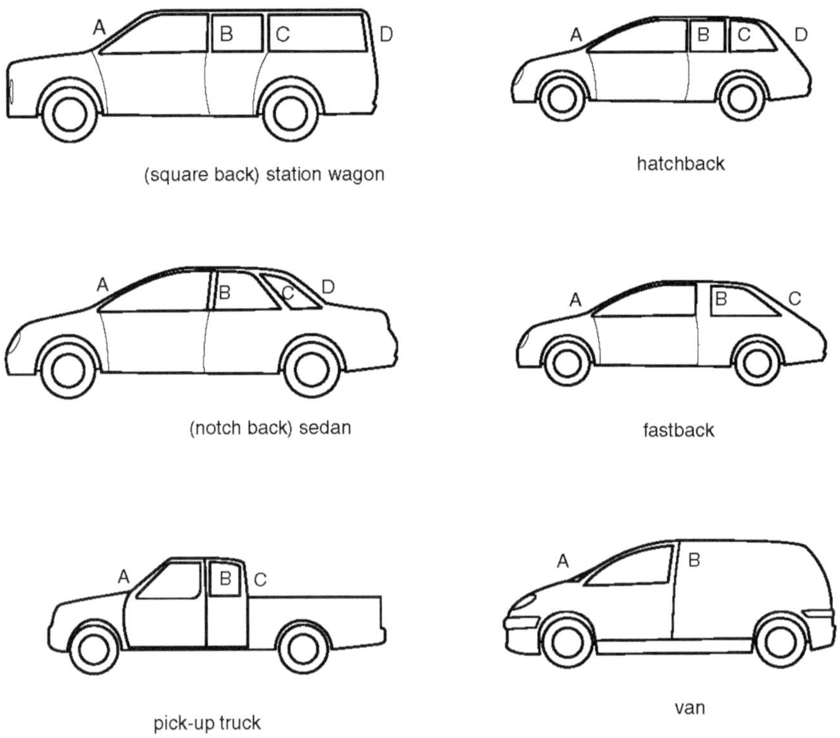

Fig. 6.9 Types of rear section

Two trailing vortices are shed at the C- (and, where present, D-) pillars. These vortices induce a downwash over the rear of the car from the rear rooftop to the trunk bottom. The radius of the roof rear-window transition region, the C-pillar design, and the aerodynamic design of the trunk all affect the lift and drag contribution of the vehicle rear section.

Both the drag and lift rise with the rear window slant angle, ϕ, up to a critical ϕ found to be around 50° after which the lift and the drag drop significantly [6-6]. The least-drag slant angle, ϕ_c, varies directly with the length, l_r, of the car rear, for a defined trunk length (Fig. 6.10(a)). An aerodynamic irony, an obtuse slant angle ϕ is better for drag and lift reduction than the traditional slant (Fig. 6.10(b)). But such a design has not received acceptance, probably because of its unappealing aesthetics.

Few attempts have been made to develop a relationship between the rear-window slant angle, ϕ, and the trunk top subtended angle, β. In one attempt, it is suggested that the slope for best drag reduction is at $\beta = 25°$ for a typical notchback, and $\beta < \phi = 30°$ for a fastback [6-7]. This of course is an approximation based on flat surfaces for the top and the rear window. But it serves as a reasonable starting point for design iteration. Rounding the sides and edges of the slant reduces the strength of the correlation and may even replace the slant as the dominant determinant of the drag coefficient. It is desirable to combine the benefits of rounding and taper, but this is a challenge because the rounding blurs the flow separation point that would have been near-definite with flat-surface sharp-edges [6-8].

Comparative dimensions of the hood, passenger compartment, and trunk have cumulative contributions to the overall vehicle lift. For example, an increase in the length of the passenger compartment tends to result in a reduction in lift—a favorable contribution. A reduction in the underside angle of incidence also results in a lower lift. A high-nose vehicle has a lower lift contribution than one with the nose close to the ground. In a low-nose vehicle, the flow, being close to the ground, is trapped in a dominantly viscous layer between the underbody and the ground. Part of it flows sideways toward the lower pressure regions on the sides of the vehicle, and increases the strength of the trailing vortices and the aerodynamic lift. The secondary flow interactions between the underbody and the ground create additional lift under the vehicle in the passenger region.

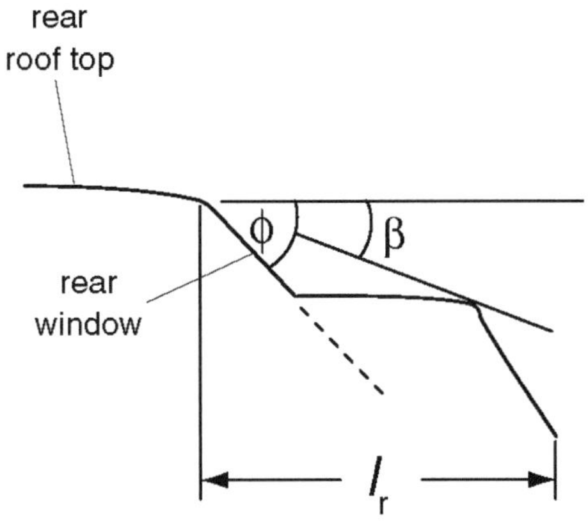

Fig. 6.10(a) Rear slant angles, ϕ and β

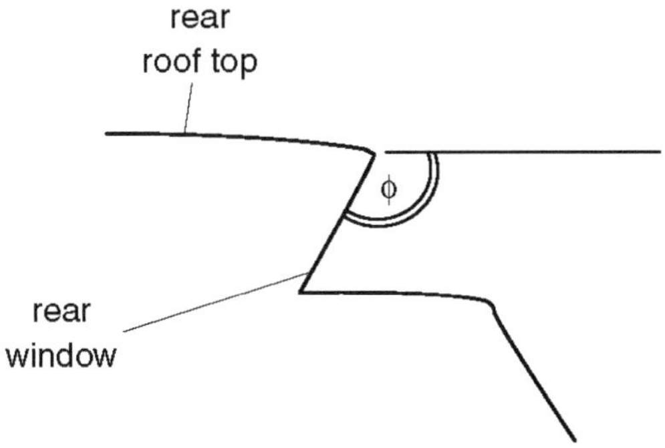

Fig. 6.10(b) Rear window slant >90°

6.3 Downforce Devices

A spoiler, common in sports cars, is a negative lift device. It reduces the lift by slowing down the flow over its upper surface. A negative lift wing is the most common type of spoiler.

When lifting devices are used, it is important to place them in the proper location at the rear of the vehicle, or they may turn out to negate the very effect for which they have been incorporated [9]. For example, a spoiler on the vehicle rear roof only adds to the lift.

When the flow pattern on the rear of the vehicle is changed (by a spoiler), the longitudinal distribution of the lift force between the axles is altered in such a way that the lift on the rear axle is reduced. With little or no influence on the front axle, the overall lift is therefore reduced, resulting in an improved stability (Fig. 6.11).

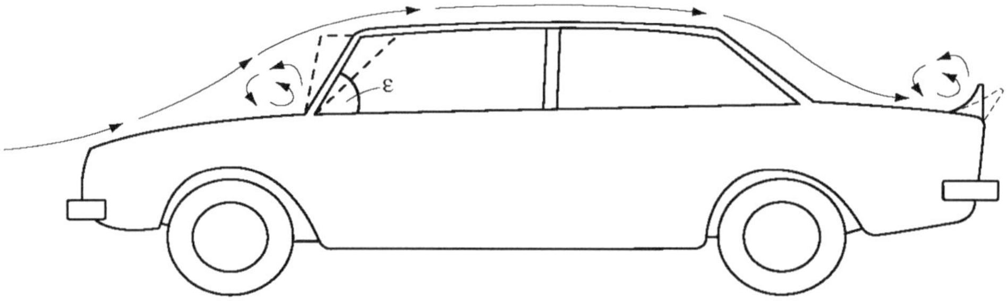

Fig. 6.11 Windscreen and rear spoiler serving as stabilizing components

6.4 Underside

The underbody air flow is a function of the clearance between the ground and the vehicle underside, the roughness of the vehicle underside, and the curvature of the underside.

The goal is to increase the flow between the vehicle underside and the ground so as to achieve downforce under the vehicle. A half-venturi longitudinal section is ideal (Fig. 6.12). The center of gravity (c.g.) can be kept low by incorporating grooved channels in the vehicle underbody. Such a design helps to

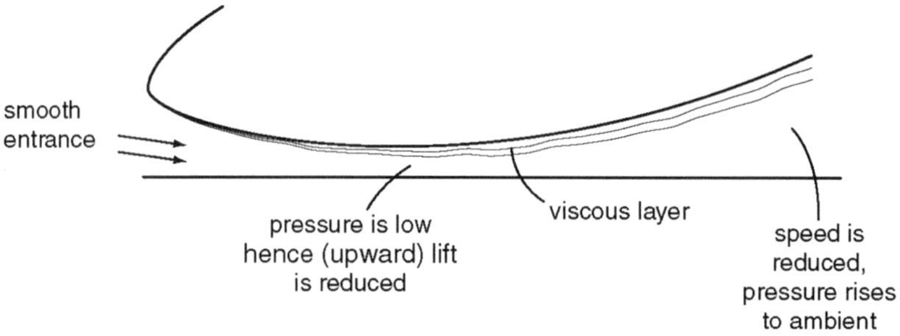

Fig. 6.12 Half-venturi (ideal) underbody

reduce the pressure and the consequential aerodynamic lift on the underbody. Underbody panels help to reduce the lift tendency brought about by the typical rough underside of vehicles. The underbody panel should also be designed with adequate lateral curvature so that the higher pressure air flow from the vehicle underside can flow toward the sides of the vehicle where the local pressure is lower than in the underbody. This lateral flow may add to the vehicle drag, but when safety is the primary consideration (as it should be), the reduction of aerodynamic lift is more of a concern than the reduction of aerodynamic drag.

The three extreme designs of flow modifiers of the vehicle underbody are front seal, total underbody seal, and rear seal (Fig. 6.13). When the vehicle front is sealed, part of the front region is still opened up to the flow around the vehicle. The underside pressure is relieved and may even become a prevailing negative pressure, even when the underside clearance is below the critical value. When the whole vehicle underside is sealed, the stagnation pressure at the nose of the car has neither straight underground nor sideways escape, and there is no opportunity to accelerate the flow under the car. These two cumulative effects result in a relatively higher lift on the vehicle. When the rear side is covered, the flow coming from the leading edge hits a wall. Although some of it is deflected sideways, the flow is predominantly deflected up against the underside of the vehicle, resulting in lift and vehicle instability, particularly at the rear of the vehicle.

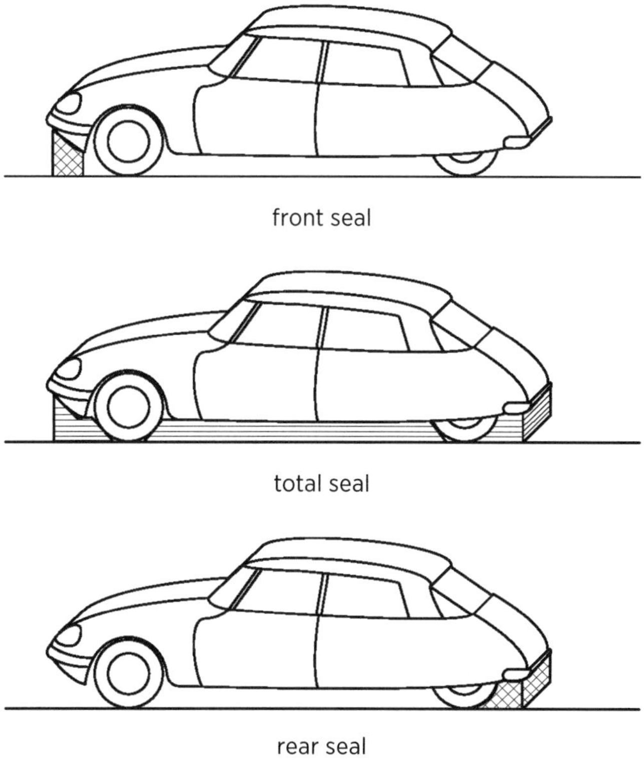

Fig. 6.13 Underbody sealing

Any styling that restricts sideways air flow—side fins, rear lamps, elevated headlamps—creates an increase in the aerodynamic lift. The opening under the hood also adds to aerodynamic lift due to a strong dynamic pressure and the inevitable flow separation at the structural members under the hood (see later Fig. 9.1).

6.5 Vehicle Height

The height of the vehicle affects the location of the vehicle's c.g. The lower the c.g., the more stable is the vehicle. For this reason and for passenger comfort, the width-to-length ratio of a car is higher than its height-to-length ratio. For trucks, which are primarily designed for cargo transport and less so for comfort, the height-to-length ratio is greater than the width-to-length ratio. The higher the car roof, the greater is the contribution to drag.

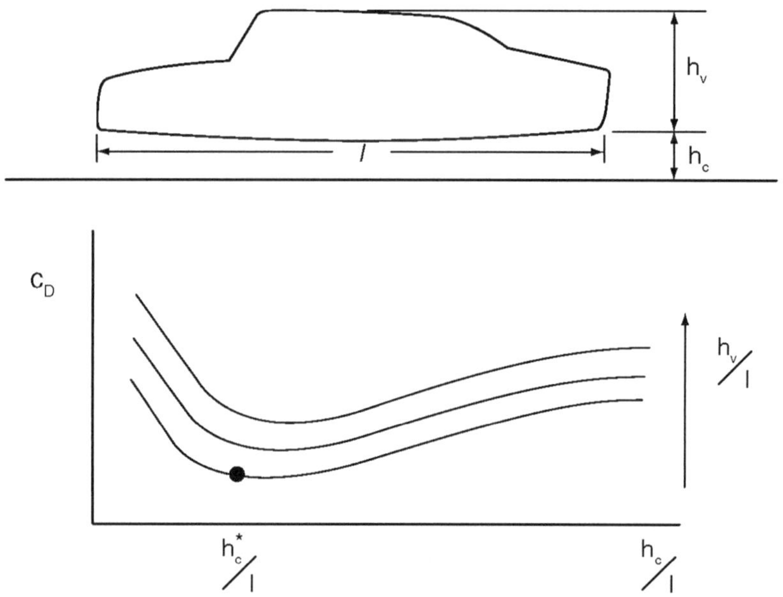

Fig. 6.14 Variation of drag coefficient with vehicle height

Figures 6.15(a) and (b) depict the respective center pressure distributions on two cars considered to be at the extremes of aerodynamic profile, with Fig. 6.15(b) being the aerodynamically favorable design. Understandably, aerodynamic design is not a primary concern in the development of some vehicles. It is, however, important to have the tools to design for drag reduction, with the accompanying benefit of fuel efficiency.

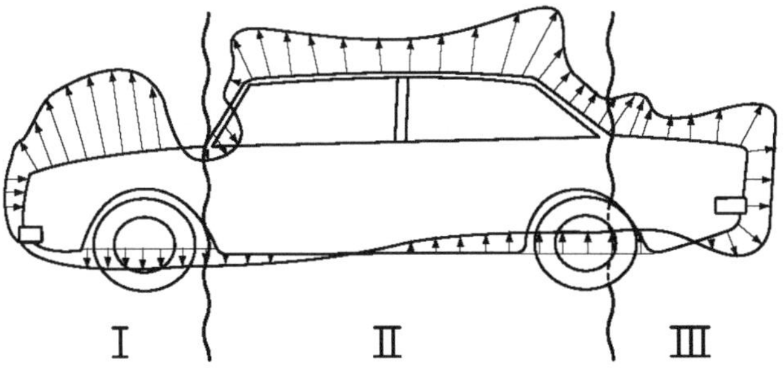

Fig. 6.15(a) Non-aerodynamic car

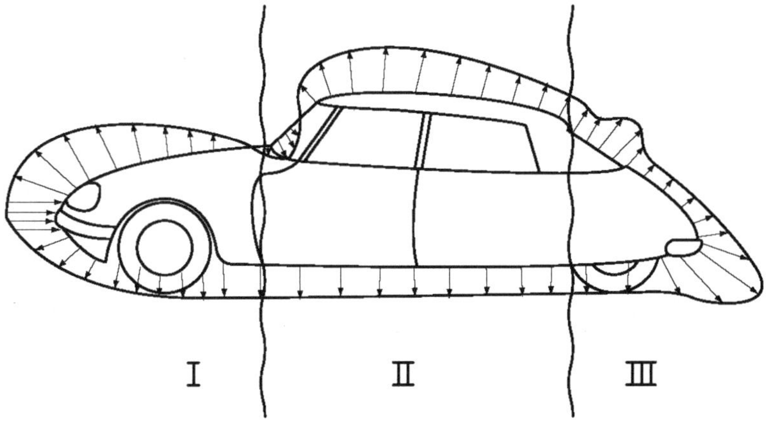

Fig. 6.15(b) Aerodynamic car

Summary

The surfaces of a car are typically of a 3-D curvature, with most of the variation in surface air flow taking place along the vehicle's longitudinal axis. The body shape of a car can be divided into three parts—front, mid, and rear—by a plane normal to the general direction of motion (i.e., a plane normal to the longitudinal axis). The front part includes the entire vehicle front up to the base of the A-pillar. The second part is essentially the cabin—from the base of the A-pillar to the bottom of the rear window. The last part covers the rest of the car—the bottom of the rear window to the rear bumper. The largest of the three cross sections is that taken around the A-pillar, separating the cabin from the hood assembly.

Increased curvature of the hood tends to move the separation point downstream. Likewise, increased curvature of the windscreen creates the beneficial tendency of moving the reattachment point to the lower end of the windscreen. Increased camber of the hood or of the windscreen will cause a delay in separation and in reattachment. Side flow restrictions such as headlamps and side trimmings are contributors to early flow separation on the vehicle. For the best results, increase the hood camber, reduce the windscreen camber to minimal shape for reattachment, and modify the sides of the cambered windscreen for added curvature for the side flow. Keep the vehicle nose low to just above a nose height that corresponds to the critical underbody clearance. Because a low vehicle nose usually requires lowering the radiator, a cross-flow radiator would be a better fit than a down-flow one for the architectural and aerodynamic constraints of an optimum car leading edge because the cross-flow radiator has a relatively high width-to-height ratio. Also, the posts and the door handles should blend with the car body. A low car roof is most desirable for drag reduction.

Key Terms

Camber: Vehicle camber is a measure of the arcing between defined points on a straight line in the vehicle lateral axis.

Curvature: Vehicle curvature is a measure of the arcing between defined points on a straight line in the vehicle longitudinal axis.

Interference: A result of the interaction of the main body drag with the drag due to a protruding surface; it is typically determined from a wind tunnel test.

Review Questions

1. What is the difference between camber and curvature?
2. How does the hood curvature affect the aerodynamics of the airflow over the hood?
3. How do the curvature and the camber of the windscreen influence the separation of air flow over a vehicle?

Projects

Perform a wind tunnel test in which the only variable is the windscreen height (hence the roof height). Compare the aerodynamic forces at the base of the windscreen with the same parameters at the top end of the windscreen. Can you, from this experiment, predict the contribution to the vehicle roof drag due to this windshield variable geometry?

For reference, pick a hood-windscreen camber-curvature combination (four variables). Determine the drag effect due to respective changes in (a) hood camber, (b) hood curvature, (c) windshield camber, and (d) windshield curvature.

References

1. Hoerner, S. H. *Fluid Dynamic Drag*, Hoerner 1951.
2. Prandtl, L., and Tietjens, O. C., *Fundamentals of Hydro and Aeromechanics*, Dover, 1935.
3. Streeter, V. L., *Handbook of Fluid Dynamics*, McGraw Hill, 1961.
4. Carr, G. W., "Potential for Aerodynamic Drag Reduction in Car Design," Dorgham, M. A., ed., Impact of Aerodynamics on Vehicle Design, Technological Advances in Vehicle Design Series, SP3, Interscience Enterprises Ltd., 1983.
5. Hucho, W. H., "Wind Tunnels for Automobile Aerodynamics," *Aerodynamics of Road Vehicles*, Butterworks-Heinemann, 1990.
6. Morel, T., "The Effect of Base Slant on the Flow Pattern and Drag of Three-Dimensional Bodies with Blunt Ends," Sovran, G., Morel, T., and Mason, W. J., eds., *Aerodynamic Drag Mechanism of Bluff Bodies and Road Vehicles*, Plenum Press, 1978, pp.191–226.

Chapter 7
Trucks, Trailers, and Buses

Until recently, there has been very little aerodynamic consideration in sizing and shaping large trucks and buses. Their shapes and sizes are primarily determined by their intended functions, which is payload transportation. Aerodynamic consideration has traditionally been an afterthought. For small and mid-size trucks, however, there has often been some aerodynamic consideration in the design, although payload is still the primary design factor. In Fig. 7.1, if H is the vehicle height, L the length, and B the vehicle width, then

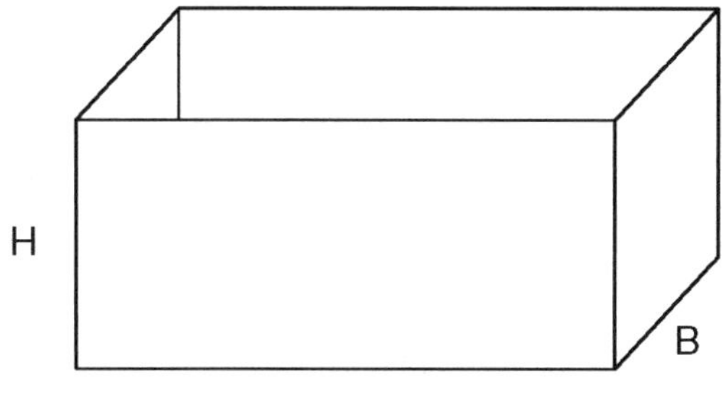

Fig. 7.1 Vehicle boxed volume

$$1 > \frac{B}{L} > \frac{H}{L} \text{ for cars} \tag{7.1}$$

and

$$1 > \frac{H}{L} > \frac{B}{L} \text{ for trucks and buses,} \tag{7.2}$$

that is, cars are wider than they are high, whereas trucks and buses are higher than they are wide.

There are some mid-size trucks that are built or modified such that they are wider than they are high, but the above relationship generally holds true. That is understandable, since passenger comfort is the overriding factor in car design, whereas cargo capacity is the mission specification of a truck.

In order to optimize the efforts put into drag reduction, it is worthwhile to consider the drag distribution in these vehicles. Typically, for tractor trailers, about 40% of the drag comes from the vehicle front section, the rear accounting for 20%. The tractor-trailer gap accounts for up to 20%, the underbody 15%, and drag due to vehicle skin friction is about 5%. For a full van or truck, 60% of the drag is attributed to the front section, 20% to the rear, 15% due to the underbody, and 5% due to skin friction [7-1].

7.1 Light Trucks, Minivans, and Sport Utility Vehicles (SUVs)

Light trucks were initially designed for small payload transportation. They have since evolved into "trucars"—a truck that is reversibly transformable into a car. Therefore, there are several car design considerations that are incorporated into the design of light trucks. Light trucks and SUVs generally have a higher ground clearance than minivans (Fig. 7.2).

Fig. 7.2 Minivan, truck, and SUV

They are therefore more likely to have a higher underbody drag. The high ground clearance and the four-wheel-drive systems typical of trucks and SUVs enhance their towing capability, especially at low speeds and in off-road uses.

(At high speeds, i.e., above 50 mph, the specific fuel rate consumption increases rapidly, with trucks consuming significantly more fuel than SUVs.) Because of the higher ground clearance, trucks have a higher center of gravity (c.g.) location, which makes them more susceptible to lateral instability.

The front section of a minivan is also more aerodynamic than that of a truck or an SUV. These differences underline the primary mission specification of each vehicle—minivans to carry passengers, trucks to carry cargo, and SUVs to travel rugged terrains. Yet, a better understanding of the flow structures around light trucks can result in a substantial improvement in their aerodynamic design [7-2].

7.1.1 Front Design

With what seems to be an irreversible rise in the cost of fuel, the recent light truck and SUV design trend has been to steer toward streamlining the vehicle leading edge. This alone can reduce the drag by about 0.1 (from $C_D = 0.39$ to $C_D < 0.3$). This is very significant when we consider that a ΔC_D of 0.01 amounts to a saving of about 0.1 mpg for trucks and about 0.14 mpg for SUVs.

The higher cooling needs in SUVs and particularly in trucks are an inevitable obstacle to the leading-edge profile of trucks approaching that of a car. Even the aerodynamic improvement that has been made comes with a price: the hood panels have become more prone to deformation and to vibration under aerodynamic loads. With ever more stringent pedestrian crash standards to provide safer front end designs to minimize head and leg injuries, the trend has become toward flexible hoods that crush more readily than typical current designs for increased absorption of crash energy. Such designs, because of the reduction in rigidity, increase the vibration of the hood as the vehicle speed increases. High-speed wakes and wind gusts also contribute to hood vibration and deformation [7-3, 7-4]. This problem of hood deformation and vibration under varying highway driving conditions is being aggressively tackled with simulation tools that enable pre-build studies of the fluid-structure interaction in a proposed vehicle design [7-5].

7.1.2 Truck Tailgate

One of the contributors to drag in trucks is the truck tailgate. It is blunt and of a rectangular planform, located behind the cab. The design alone is a drag contributor because of the more or less blunt corners. Additionally there are tailgate on and off conditions. When the tailgate is on (Fig. 7.3(a)), part of the air flow passing over the cabin initially falls into the bed, and builds an air wall. The air wall creates the same effect as covering the bed with a flat panel. The subsequent air flow travels over this artificial wall and sheds off at the rear of the tail gate. In Fig. 7.3(b), the tailgate is left open, flat level with the bed floor. In this case, there is no endplate, as in Fig. 7.3(a), to contain the flow; hence no air wall can be formed. The added length of the gate and the resulting downwash contribute to the drag at the rear (a disadvantage) but add to the rear downforce (a benefit).

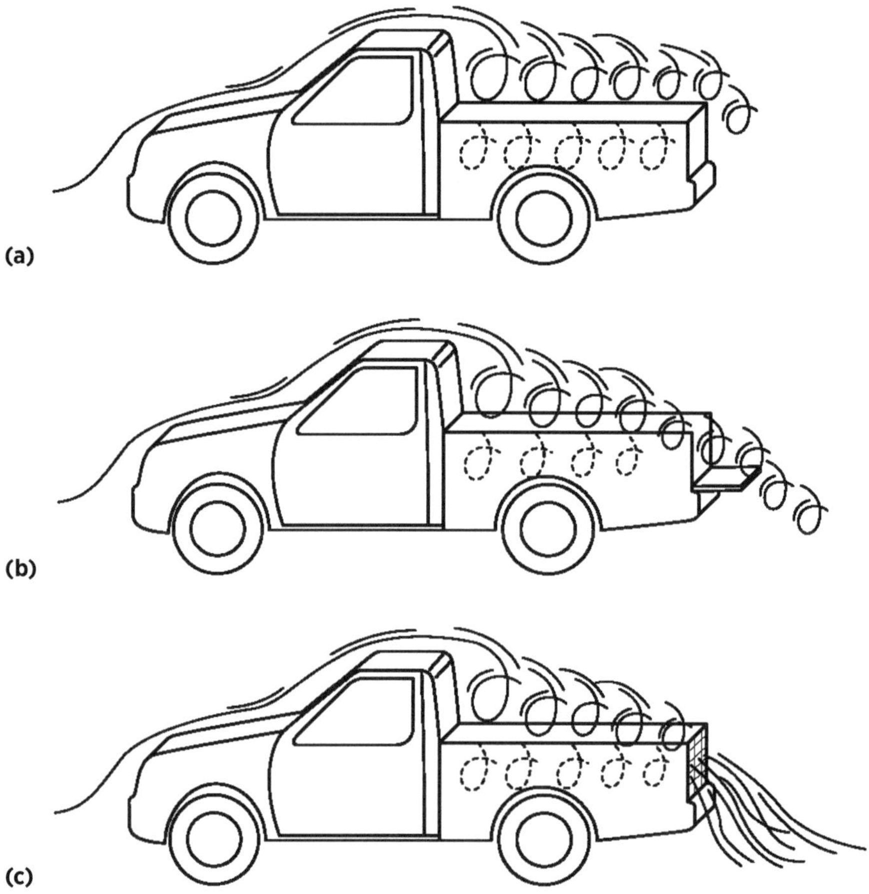

Fig. 7.3 Truck rear guard

In Fig. 7.3(c), a tailgate net replaces the tailgate. Because it is lighter than a tailgate, it may provide a better fuel efficiency. But, aerodynamically, it is a drag contributor. At low speeds, the air flow is close to Fig. 7.3(b), but with a greater drag due to multiple flow separation on the net. At higher speeds, the flow is turbulent, and forms a porous block at the net. The result is a flow that approaches that of Fig. 7.3(a), but with more drag. In the three tailgate configurations considered, the order of drag contribution from the least to the highest is tailgate-on, tailgate-flat, drag net-on. Two additional configurations, which are not common, are tailgate-off and flat bed configurations. When these two configurations are included, the rising order of drag contribution is flat bed, tailgate-on, tailgate-flat, tailgate-off, drag net-on.

Other factors that influence the drag for the varied tailgate configurations are speed and yaw angle. In general, higher drags are generated at yaw angles. The tailgate-on configuration appears to be more sensitive to yaw angle variation. Beyond some critical value (about 15°), the yaw has negligible effect on the drag contribution of any of any of the tailgate configurations [7-6].

Notwithstanding that trucks are commonly used for rugged activities, tailgate noise reduction is a desirable goal in truck design, and efforts are being made to optimally constrain tailgates to achieve this design goal. One of the findings in this effort is that the diagonal distortion of the tailgate opening must be minimized in order to prevent tailgate chucking—a type of rattle noise caused by the contact between the latch and the striker in rugged driving conditions [7-7]. A traditional approach has been to add wedges in the gate opening. Although this approach addresses tailgate chucking, it introduces another challenge—increased sensitivity to induced boom from the tailgate [7-8]. This is critical because boom, a structurally generated noise by body panels, causes passenger discomfort (as discussed in chapter 2). An isolated striker has been found to function adequately in preventing boom. An isolated striker is a rubberlike dampening material embedded between two sheet metal layers.

7.2 Medium Trucks

Medium trucks are those with gross weight ranging from 14,000 to 26,000 pounds (6351 to 11793 kg). They are the largest category of commercial trucks. The front of most medium trucks is mainly a bluff body, although there is a recent trend to present an aerodynamic front. They have a higher ground clearance than light trucks and hence a lower downforce.

Enhanced traction, rather than drag reduction, is the primary goal in medium truck design. Yet an aerodynamic leading-edge design would reduce the aerodynamic drag and contribute to vehicle fuel efficiency, albeit at the expense of a longer vehicle front section.

7.3 Large Trucks and Buses

The architectural difference between trucks and buses is due to the difference in payload management—people for buses and cargo in trucks. Because in this section trucks and buses will be treated as large vehicles, it is important to itemize their differences. The intrarelational dimensions of a bus are presented in Fig. 7.4.

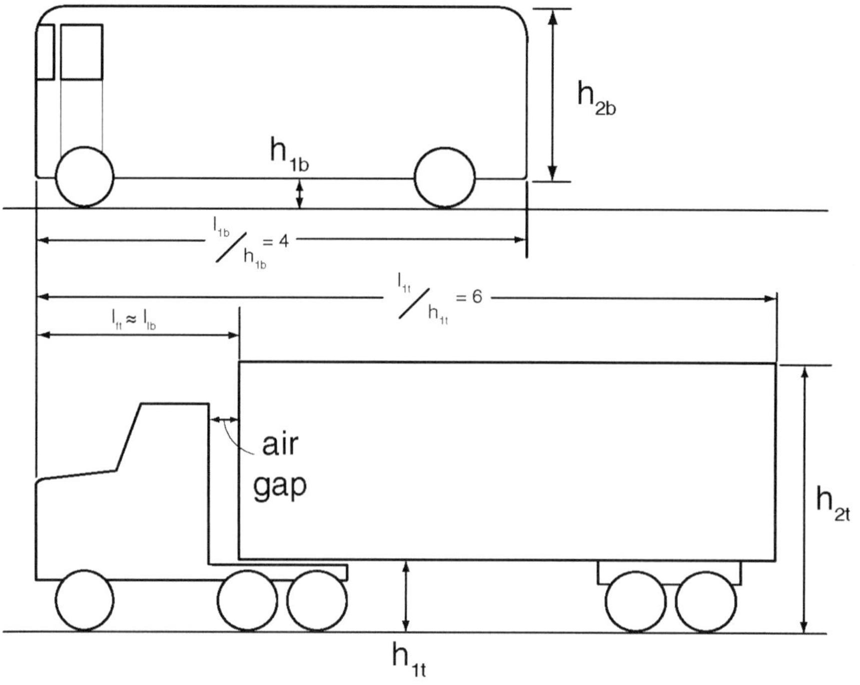

Fig. 7.4 Truck and bus: relational dimensioning

Also, the following relations hold in general:

$$\begin{gathered} w_b = w_t \\ h_{1_t} = 3h_{1_b} \\ h_{1_t} = d = \sqrt{\frac{4A}{\pi}} \end{gathered} \tag{7.3}$$

where

d is the effective diameter, and

A is the frontal projection area [7-9].

7.3.1 Large Trucks

In a common classification, there are two types of truck: cabin-behind-engine truck and cabin-over-engine truck. For the cabin-over-engine truck, the cabin is generally higher and of a shorter length than that in a cabin-behind-engine truck. It is typically the same width as the trailer, whereas the cabin width of the cabin-behind-engine truck is smaller than the width of the trailer.

The practice in tractor-trailer spacing is such that the air gap is 40–80% of the vehicle effective diameter. The wider the gap, the more the aerodynamic drag incurred up to the critical gap, beyond which the tractor and the trailer become essentially decoupled. Wind tunnel experiments have also been used to verify that there is a relationship between tractor-trailer gap and vehicle drag. At zero yaw angle, the critical tractor-trailer gap, $G_{critical}$, is given by the relationship

$$G_{critical}/\sqrt{A_g} = 0.5 \tag{7.4}$$

where A_g is the gap cross sectional area (normal to the longitudinal flow, typically the trailer width). At the critical width, there is a jump in the drag, with a large proportion of the drag estimated to be due to the trailer. With increasing gap, the tractor and trailer become decoupled, and the drag steadily rises to a terminal value [7-10].

Tractor trailers have been tremendously improved from a more or less flat front to the current aerodynamic front assembly (Fig. 7.5). Most of the aerodynamic improvement is focused on the front section. The roof deflectors and side extensions (Fig. 7.5(c)) tremendously improve the aerodynamic characteristics of the flow. Even simple modifications, such as the addition of fences to the front of the trailer (Fig. 7.6), make a noticeable improvement to the aerodynamics of flow [7-11]. Interestingly, a flat-plate fence yields a much better result than an airfoil-shaped fence.

(a)

(b)

(c)

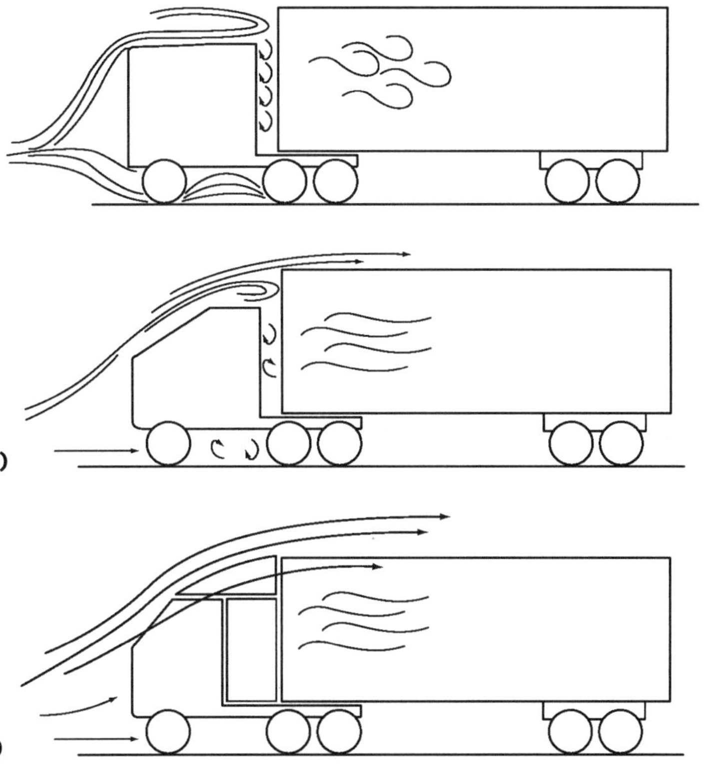

Fig. 7.5 Progression in tractor-trailer design

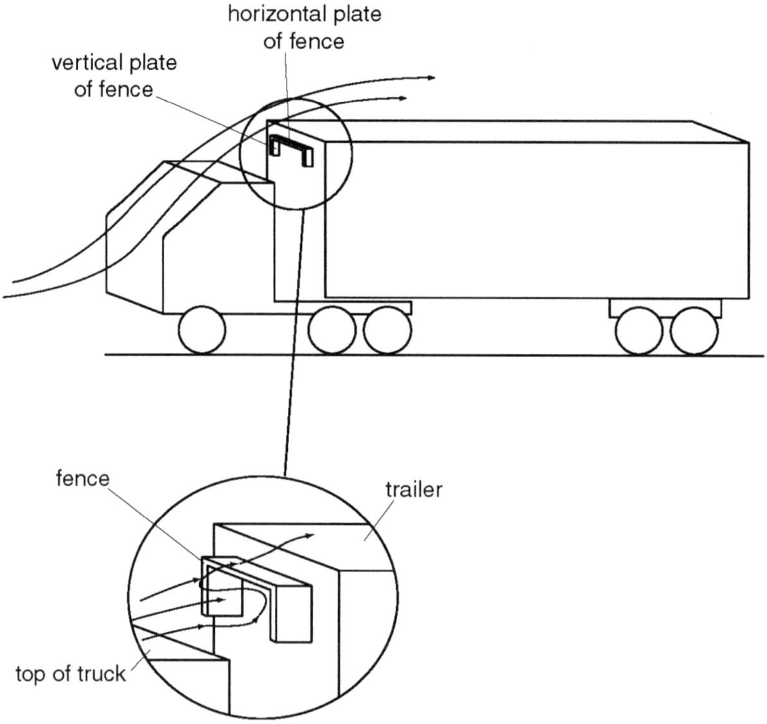

Fig. 7.6 A trailer with fences installed

Concern over aerodynamic lift is not of great importance in large trucks because of the sufficiency of the sheer weight of the truck to overcome the lift. The greatest concern is the aerodynamic drag, and the larger the vehicle, the greater is the aerodynamic drag. For tandem tractors of the same total capacity as a single tractor, there is a substantial aerodynamic drag increase due to the gap between the two carts in the tandem design (Fig. 7.7).

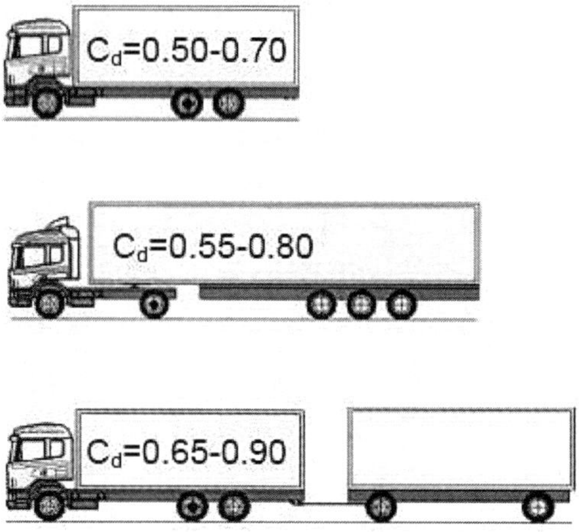

Fig. 7.7 Drag comparison for truck lengths and designs

It has been shown that an easy way to decrease the specific aerodynamic drag of a tractor-trailer is to add one or two more trailers. For example, when a second trailer is added, the capacity of the tractor trailer is doubled but the vehicle weight is less than double, and the aerodynamic drag increases by only about 40% [7-12]. The specific aerodynamic drag is therefore 70% of the single-trailer tractor—a 30% reduction.

7.3.2 Buses

Because of the continuous uniform surfaces of a bus, it is perhaps easier to predict the flow pattern and to modify the flow on a bus.

Consider the top of the school bus as a long, flat plate 40 ft long, 8 ft wide. The bus is traveling at 60 mph under standard temperature and pressure conditions. If the critical Reynolds number is 10^6, how far from the front of the bus roof will transition begin to take place from laminar to turbulent flow? How much skin friction drag is attributable to the top of the bus?

$$\mathrm{Re} = \frac{\mathrm{v}x}{\nu}$$

$$x = \frac{\mathrm{Re}(\nu)}{\mathrm{v}} = \frac{10^6 \times (0.0001567)}{88}$$

$$= 1.78068 \text{ ft}$$

Transition begins just under 2 ft from the leading edge. That means over 90% of the flow on the roof of the school bus is turbulent flow.

Before we go on to determine the skin friction, let's consider the school bus in a 20 mph restricted zone. In this case, transition would take place 5.34 ft from the leading edge, consistent with the expectation of a delayed (and sometimes evaded) turbulent flow.

The friction drag, D_f, is defined as

$$D_f = qAC_f$$

where

C_f is the skin friction;

q is the dynamic pressure ($\frac{1}{2}\rho \mathrm{v}^2$); and

A is the surface area of the top of the bus.

The laminar and turbulent skin friction drag coefficients are, respectively

$$C_{f_{laminar}} = \frac{(1.328)}{\sqrt{\mathrm{Re}}},\ C_{f_{turbulent}} = \frac{0.455}{(\log_{10}\mathrm{Re})^{2.58}}$$

Laminar drag is $D_{f_{laminar}} = qA_l \dfrac{(1.382)}{\sqrt{Re}}$

where A_l is the area of the laminar section of the flow.

$$q = \frac{1}{2}0.002378(88)^2 = 9.207616, \ A_l = 1.78068 \times 8 = 14.24544$$
$$D_{f_{laminar}} = 0.1742 \text{ lb}$$

To find the turbulent component, first determine what the turbulent contribution would be up to the critical length of 1.78068 ft, then subtract the same from the turbulent contribution for the entire length of the roof:

$$D_{f_{turbulent}} = qA\frac{0.455}{(\log_{10} Re)^{2.58}}.$$

$$A_1 = A_l = 1.78068(8) = 14.2454, \ Re_1 = 10^6$$
$$A_2 = 40(8) = 320, \ Re_2 = \frac{88 \times 40}{0.0001567} = 22.4633 \times 10^6$$

$$\begin{aligned} D_{f_{turbulent}} &= D_{f_{turbulent_2}} - D_{f_{turbulent_1}} \\ &= 17.1416 - 0.5864 \\ &= 16.5552 \end{aligned}$$

$$\begin{aligned} D_f &= D_{f_{laminar}} + D_{f_{turbulent}} \\ &= 0.1742 + 16.5552 \\ &= 16.7294 \text{ lb} \end{aligned}$$

With such information, a bus designer has a fairly good idea of the locations to make modifications, such as delaying the onset of turbulence. The flow does not change straight from laminar to turbulent as depicted. It goes through a transition region that can be estimated and used to refine the result shown [7-13, 7-14]. The presented approach suffices to guide the designer and the aerodynamicist close to the location where modification would be effective for drag reduction.

7.3.2.1 Articulated (Accordion) Buses

An improvement in aerodynamic drag similar to that of a multitrailer tractor is obtained with an accordion bus over a single-cabin bus. Due to the articulation, the length of the accordion bus as well as its capacity is increased by 50% but the drag is increased by only 9% [7-12]. This results in a 27% reduction in specific aerodynamic drag.

7.3.2.2 Double-Decker Buses

A double-decker bus is a high, bluff body. Efforts have been made to reduce the tremendous leading-edge drag inherent in the design. A common aerodynamic improvement is to pull back the leading edge of the top deck, morphing the profile from the roof at the top deck to the lower fascia at the vehicle underbody. Although this results in improved vehicle aerodynamics, it reduces the number of passengers in the top deck—a commercially unappealing situation.

7.4 Side Skirts

Trailer side skirts are flat plates suspended on either side of the trailer underbody (Fig 7.8). Between 60 and 70% of the tractor-trailer drag coefficient is due to the tractor, and 15 to 20% is due to the trailer cargo box and to the trailer wheel assembly, respectively. The flow in the underside of a trailer is such that a large recirculation zone separates from the tractor drive wheels and the underbody and extends rearward. The skirts provide a surface to which the underbody flow can reattach, thereby reducing the recirculation zone. The longer the skirt, the greater will be the flow attachment, and the lower will be the drag coefficient. In particular, the skirts, when properly positioned, reduce the vehicle drag by shielding the trailer wheel assembly from the high-speed, cross-stream flow that passes beneath the trailer when the vehicle is operated within cross wind. Side skirts appropriately mounted on the underside of heavy-duty truck trailers can provide a reduction of up to 15% in fuel consumption [7-15].

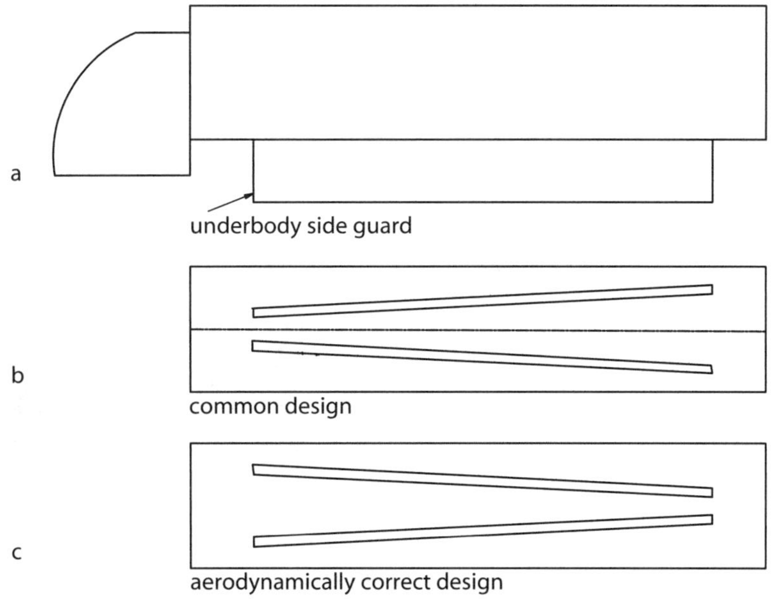

Fig. 7.8 Trailer underbody side skirts

Their design reduces the ground clearance of the vehicle. Therefore, they are often damaged when the vehicle passes over railroad crossings or when the vehicle is backed into sunken loading docks. They also collect a substantial amount of ice, which results in an increase in the vehicle gross weight. The consequences include increased specific fuel consumption and possible excess weight penalty at truck weigh stations. Therefore, as in most equipment design, the benefits and the disadvantages need be considered for an optimum design. For example, from an aerodynamic perspective, Fig. 7.8(c) is the best design. The progressive reduction in cross-sectional area results in an increased velocity downstream and a correspondingly reduced pressure. A reduced underbody pressure reduces the lift and increases the effective downforce. The result is better vehicle stability. In Fig. 7.8(a), the skirts are straight. There is no Bernoulli-based contribution to the downforce as found in Fig. 7.8(c). The least aerodynamically favorably design is Fig. 7.8(b). The skirts are divergent, resulting in progressively increased underbody pressure as the cross-sectional area widens. This contributes to vehicle lift and instability. Yet Fig. 7.8(b) is the common design found on trailers, for reasons that apparently overshadow the aerodynamic consideration. One such reason is the location of the trailer release jack (Fig. 7.9). To design for best aerodynamic consideration, one would either have to enclose the jack within the skirts or start the skirts after the jack. The first option can be a nightmare each time the operator wants to disengage the truck from the trailer. The second option fails to take full advantage of the skirts. Also, the trucks are generally heavy enough, even without cargo, that there is hardly any concern about uplift. Therefore, the disadvantage of the divergent skirts is overcome by the shear weight of the trucks that incorporate side skirts. Such factors as these result in an optimum design that may include features that when viewed in isolation, appear to be an antithesis to the design goal.

Fig. 7.9 Location of trailer release jack relative to side skirt

7.5 Truck-Car Interaction—A Further Treatment

The aerodynamic effect that a vehicle in traffic has on other vehicles is experienced through local pressure fields when overtaking and through wake interaction downstream. In chapter 5, a passenger car passing a truck is treated for the simple case of the car and the truck moving in a head wind (see chapter 5, Figs. 5.17 and 5.18). The results are different in cross wind and in tail wind. Tail wind is aerodynamically beneficial as it offsets the drag on the vehicle. The concerns about tail wind condition are braking and rear vehicle lift. Of greater interest is the cross wind flow in car-truck flow interaction. Consider both the car and truck to be heading north, with the car in the passing lane—typically the left lane—and overtaking the truck. We will describe two headwind conditions: one from the northeast and the other from the northwest (Fig. 7.10). There are three phases in each condition: car approaching truck zone, a car in the truck zone, and a car leaving the truck zone. In truck-car interaction, because the passenger car's shape typically generates lift, the wake structure is dominated by trailing vortices, whereas the wakes from trucks are of bluff, nonlifting nature with transverse velocity components that strongly affect smaller vehicles in the truck lateral environment, as will be shown in the second phase—car in truck zone—of the truck-car interaction.

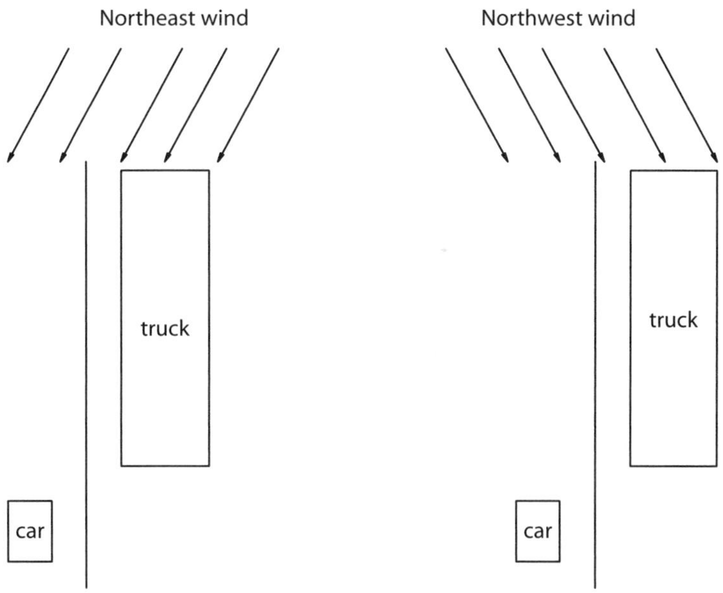

Fig. 7.10 Headwinds from northeast and northwest on north-bound vehicles

7.5.1 Northeast Headwind Condition

As the car approaches the truck zone, the modified northeast wind and the transverse component of the truck sidewash tend to push the car such that it heads in a northwesterly direction (Fig. 7.11). The car driver must compensate by steering toward the truck. In the truck zone, the airflow is tunneled primarily due to the wall effect of the trailer panel. The longitudinal component of the flow is therefore dominant. As a result, the car driver does not need to compensate as much in order to keep a straight path. As the car leaves the truck zone, it experiences a phenomenon similar to that at its approach to the truck zone. One would expect the side force to be more pronounced at this stage than at the approach, but that is not the case. This is perhaps due to the truck leading-edge flow stagnation.

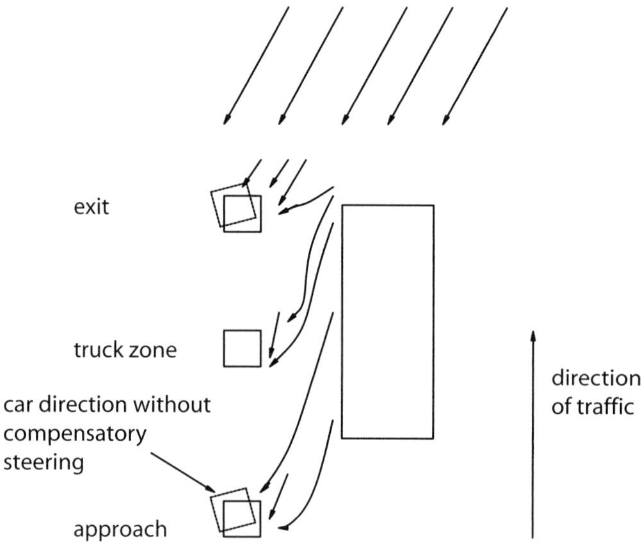

Fig. 7.11 Northeast headwind condition

7.5.2 Northwest Headwind Condition

When the headwind is from the northwest, the car is pushed toward the truck by the gust (Fig. 7.12). But, unlike the case of the northeast wind, the transverse component of the truck sidewash reduces the effect of the headwind such that the driver requires only a slight effort to steer the car from the truck and

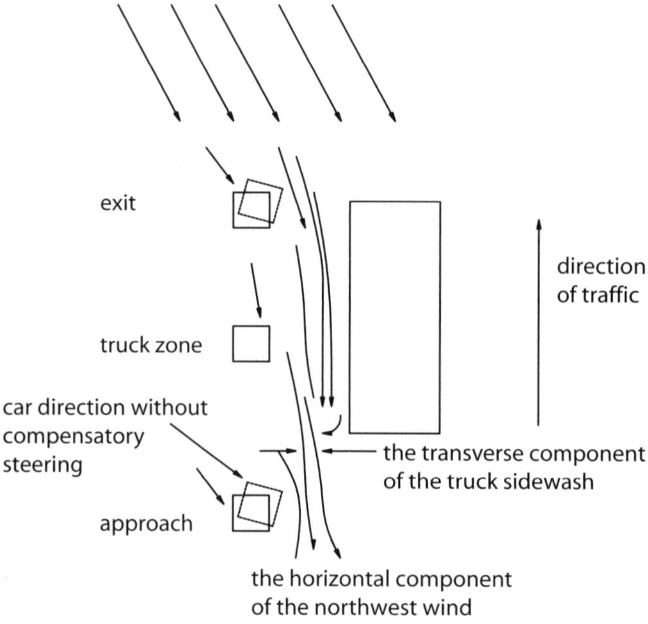

Fig. 7.12 Northwest headwind condition

have it head straight. In the truck zone, there is a consistent air wall parallel to the trailer panel, just as in the case of the northeast headwind. Finally, on exiting the truck zone, the headwind pushes the car toward the truck. This, the pressure drop between the car and the truck increases the draw of the car toward the truck. The driver therefore has to put more effort into steering the car away from the truck than would be required at the entry into the truck zone.

7.6 Applications of Computational Aerodynamics to Large Ground Vehicles

It is common knowledge that at speeds above 50 mph, the aerodynamic drag dominates the other resistant forces. Seeing that trucks average a speed of 65 mph, it is expected that there are benefits in improving their aerodynamic design. One of the areas of drag reduction is in the vehicle underbody. Vehicle underbody drag reduction is implemented in cars and in small trucks with results of ground simulations, inclusively [7-16, 7-17]. The weight and size of large vehicles make it currently impractical to provide such ground simulations for large vehicle full-scale tests. Computational methods become particularly useful in such a situation, and findings from computational simulations have contributed to improvements in wheel housing and lower body design for large vehicles [7-18]. Methods such as unsteady Reynolds averaged Navier-Stokes (URANS) simulation and large eddy simulation (LES) have been used successfully in predicting and in fine-tuning complex flows around large vehicles, especially at low Reynolds numbers.

Summary

Trucks and buses, unlike cars, are higher than they are wide. Trucks and buses are typically designed for payload. Aerodynamic profiling for drag reduction has for some time been an afterthought in the design of buses and especially trucks.

Most of the aerodynamic drag in trucks and buses comes from the vehicle front section.

Relatively little emphasis is placed on lift in large trucks and in bus design because the sheer weight of the vehicle readily overcomes any lift force that may be generated (most of which is from the vehicle underbody).

Recently, add-ons such as side skirts have been introduced to reduce the aerodynamic drag contribution from the vehicle underside.

Key Terms

Side skirts: Plates mounted on the sides of the trick underbody to reduce aerodynamic drag contribution coming from the vehicle underside.

Tandem trailer design: A design in which two separate trailers are co-hitched and the leading trailer is hitched to the tractor.

Tractor trailer spacing: The gap between the tractor and the trailer; the bigger this gap, the higher is the drag contribution.

Review Questions

1(a). Which section of a truck is responsible for most of the aerodynamic drag?

(b). Which typically has a higher ground clearance: a large truck or a tour bus?

2. Describe a side skirt. How do side skirts improve the aerodynamics of a large truck?

Application Questions

1. Which would be more economical for a family of five to purchase from a fuel efficiency perspective: an SUV or a van of the same cabin room? Assume that both cost the same at the time of purchase.
2. A tandem trailer weighs more and incurs more drag than a single trailer of the same capacity. Discuss the justifications for using a tandem trailer despite the disadvantages.

Exercises

1. Find the critical gap for a tractor trailer with a trailer width of 12 ft.
2. Find the first pass design height of a tractor trailer with a frontal projection area of 110 ft.
3. A tour bus 50 ft long and 8.5 ft wide is traveling at 65 mph in standard temperature and pressure conditions. If the critical Reynolds number is 106, how far from the front of the bus roof will transition begin to take place from laminar to turbulent flow? How much skin friction drag is attributable to the top of the bus?

Projects

1. Discuss the changes in aerodynamic characteristics of a tractor and trailer combination as it turns at increasingly tighter radiuses.
2. Compare the drag on an accordion bus with that of a single rigid counterpart of the same length.

References

1. Drollinger, R. A., "Heavy Truck Aerodynamics," SAE Paper 870001, 1987.
2. Zhu, H., and Yang, Z., "Simulation of Flow Around a Generic Pickup Truck with RSM Model," SAE Paper 2008-01-0324, 2008.
3. Duncan, B., Senthooran, S., Hendria, D., Sivakumar, P., Freed, D., Gleason, M., and Hall, D., "Multidisciplinary Aerodynamic Analysis for Production Vehicles: Application of External Flow Simulations to Aerodynamics, Aeroacoustics and Thermal Management of a Pickup Truck," SAE Paper 2007-01-0100, 2007.
4. Ramsay, T., Fredelake, A., and Stevens, K., "Correlation of a CAE Hood Deflection Prediction Method," SAE Paper 2008-01-0098, 2008.
5. Gupta, A., Gargoloff, J. I., and Duncan, B. D., "Response of a Prototype Truck Hood to Transient Aerodynamic Loading," SAE Paper 2009-01-1156, 2009.
6. Mokhtar, W. A., Britcher, C. P., and Camp, R. E., "Further Analysis of Pickup Trucks Aerodynamics," SAE Paper 2009-01-1161, 2009.
7. Kuo, E. Y., and Mehta, P. R., "The Effects of Body Joint Designs on Liftgate Chucking Performance," SAE Paper 2005-01-2541, 2005.
8. Imam, A. S., Chinta, B., and Thorpe, S., "Optimum Constraint Strategy for Liftgates," SAE Paper 2011-01-0766, 2011.
9. Mason, Jr., W. T., and Beebe, P. S., "The Drag related Flow Field Characteristics of Trucks and Buses," *Aerodynamic Drag Mechanisms of Bluff Bodies and Road Vehicles*, Sovran, G., Morel, T., and Mason, Jr., eds., Plenum Press, 1978.
10. Hammache, M., and Browand, F., "On the Aerodynamics of Tractor-trailers," The Aerodynamics of Heavy Vehicles: Trucks, Buses and Trains, United Engineering Conference, Monterey, California, Dec. 2–6, 2002, p. 185–205.

11. Modi, V. J., St. Hill, S., and Yokomizo, T., "Drag Reduction of Trucks Through Boundary Layer Control," *Journal of Wind Engineering and Industrial Aerodynamics*, 54/55, 1995, pp. 583–594.

12. Cooper, K. R., "Commercial Vehicle Aerodynamic Drag Reduction: Historical Perspective as a Guide," The Aerodynamics of Heavy Vehicles: Trucks, Buses and Trains, Monterey, California, December 2–6, 2002, pp. 9–28.

13. Dodge, R. A., and Thompson, M. J., *Fluid Mechanics*, McGraw-Hill, 1937.

14. Dommasch, D. O., Sherby, S. S., and Connolly, T. F., *Airplane Aerodynamics*, 4th ed., Pitman Publishing Corporation, 1967.

15. Ortega, J. M., and Salari, K., "Investigation of a Trailer Underbody Fairing for Heavy Vehicle Aerodynamic Drag Reduction," SAE Paper 2008-01-2601, Aerodynamics and Fuel Economy ~ Aerodynamic Devices and Methods & Technologies (SP-2219), Commercial Vehicle engineering Congress & Exhibition, Rosemont, Illinois, October 7–9, 2008.

16. Elofsson, P., and Bannister, M., "Drag Reduction Mechanisms Due to Moving Ground and Wheel Rotation in Passenger Cars," SAE Paper 2002-01-0531, 2002.

17. Waschle, A., "The Influence of Rotating Wheels on Vehicle Aerodynamics—Numerical and Experimental Investigations," SAE Paper 2007-01-0107, 2007.

18. Soderblom, D., Lofdahl, L., Elofsson, P., and Hjelm, L., "Heavy Vehicle Wheel Housing Flows—A Parametric Study," SAE Paper 2009-01-1169, 2009.

Chapter 8
Railroad Train Aerodynamics

As a train travels, it displaces the air in and around the region of its travel, thereby forming a slipstream, a boundary layer along the path of its travel. Consistent with boundary layer theory, the air at the surface of the train (also referred to as the air at the wall) travels at the speed of the train. Farther from the boundary layer, the air travels at the ambient condition. The boundary layer formed on a train is typically a turbulent boundary layer; this is true even for a slow-traveling cargo train because of the typical bluff-body design of such trains (Fig. 8.1).

Fig. 8.1 Cargo train

When the Shinkansen high-speed train began operation in Japan in 1964, it was operating at a service speed of 130 mph (210 km/h), and the aerodynamic drag was two-thirds of the total drag. In 2012, the speed was over 193 mph (300 km/h), with the goal to operate at the Maglev record speed of 361 mph (582 km/h). Air resistance, the aerodynamic drag, is significant at all train speeds. Unlike in the case of cars and trucks, where the primary source of drag is the vehicle itself, in train aerodynamics three structures contribute to the aerodynamic drag: subject train, the other-train, and the tunnel.

The subject train contribution is made up of the primary profile drag and the intercoach interference drag, just as with a truck-trailer or a fleet of cars on the highway. [Profile drag = Skin friction drag + Form drag, and Parasitic drag = Skin friction drag + Form drag + Interference drag = Profile drag + Interference drag]. The typical external contributor to drag in cars and trucks is the cross wind, and it's not always present. But because of the peculiar environment, especially of a high-speed train, the contributions due to the other-train and the tunnel are inclusive in the defined total drag of the train.

The other-train is the train going in the opposite direction. Its tracks are very close to the tracks of the subject train—about two train widths away—and therefore the subject train is within the boundary layer of the other-train. The tunnel is a reflective wall and a source of turbulence because of its limited width, which puts it in the boundary layer of the train.

The complex nature of the trio—the subject train contribution, the other-train, and the tunnel—that are the major players in the aerodynamic drag of a train makes it impractical to perform a reliable wind tunnel test, except for the narrow study of a coach design. Computational methods have proven to be the most effective tool for the aerodynamic study of trains, and the effectiveness of this method improves with more efficient numerical schemes and with advancement in hardware [8-1, 8-2].

There are many types of trains performing different functions. Trains can be classified into various categories, but perhaps the simplest grouping is by the train's essential function: transportation of passengers and haulage of cargo. Passenger trains are further classified into citywide, regional, and intercity commuter trains. Citywide and regional service trains are lowspeed, although regional ones are sometimes considered medium speed (50–80 mph).

8.1 Low-Speed Passenger Train

Low-speed passenger trains are typically used for citywide transportation. They pull between six and ten cars and travel at about 50 mph, making frequent stops along the routes. Because of the relatively low speed at which these trains travel, attention is not devoted to the aerodynamic profiling of the locomotive leading edge. It is only sufficiently rounded to prevent a bluff-body frontal face (Fig. 8.2). Beyond the locomotive, the design of the upper section of the cars is essentially the same as for high-speed trains—a square box rounded at the corners.

Fig. 8.2 Citywide commuter train

Low-speed trains have a relatively high skin friction drag, only surpassed by that of a freight train.

8.2 High-Speed Passenger Train

High-speed trains are typically passenger trains pulling between 15 and 20 cars. Earlier efforts to speed up the train focused on enhancing the power of the electric motor that drives the train. While this effort has resulted in higher power output for the train, it has also led to a higher energy loss and deterioration in the performance of the train, since the flows around the train are more disturbed due to turbulence of the increased speed [8-3]. With train speeds now exceeding 200 mph (322 km/h), such aerodynamic problems as noise, vibration, impulse wave at tunnel exit, passing of oppositely faced high-speed trains on adjacent tracks, eardrum discomfort of train passengers, among others, have come to the forefront. (As shown in Table 8.1, the record speed of trains in trial runs over a 50-year period that started in 1963 has increased from about 160 mph to 360 mph—a 200 mph jump!)

Table 8.1 Records in Trial Run of Trains

Year	Country	Train	Speed	
			mph	km/h
1963	Japan	Shinkansen	159	256
1967	France	TGV 001	198	318
1972	Japan	Shinkansen	178	286
1974	France	Aérotrain	267	430.2
1978	Japan	HSST-01	191	307.8
1979	Japan	Shinkansen	198	319
1979	Japan	ML-500R (unmanned)	321	517
1981	France	TGV	236	380
1985	West Germany	Inter City Experimental	201	324
1987	Japan	MLU001 (manned)	249	400.8
1988	West Germany	Inter City Experimental	252	406
1988	Italy	ETR 500-X	198	319
1988	West Germany	TR-06	256	412.6
1989	West Germany	TR-07	271	436
1990	France	TGV	320	515.3
1992	Japan	Shinkansen	217	350
1993	Japan	Shinkansen	264	425
1993	Germany	TR-07	280	450
1994	Japan	MLU002N	268	431
1996	Japan	Shinkansen 500	277	446
1997	Japan	MLX01	342	550
1999	Japan	MLX01	343	552
2002	Spain	AVE S-102 (Talgo 350)	225	362
2002	China	China Star	199	321
2003	China	Siemens Transrapid08	311	501
2003	Japan	MLX01	361	581
2004	South Korea	HSR-350x	219	352.4
2006	Spain	AVE S-103 (Siemens Velaro)	251	404
2007	France	V150	357	574.8
2007	Taiwan	700T series train	217	350
2008	China	CRH3	245	394.3
2010	China	CRH380AL	302	486.1
2011	China	CRH380BL	303	487.3

The design of leading edge is perhaps the most crucial in high-speed train design. There are sufficient variations in high-speed train design such that a train with an aerodynamic nose ends up contending with just about the same overall aerodynamic force as one with a bluff nose traveling at a slightly lower speed. Nonetheless, a train with a streamlined leading edge has more favorable aerodynamic characteristics for high speed (Fig. 8.3), and the gap of its performance over a bluff-nosed train increases with the train speed.

Fig. 8.3 High-speed train

8.3 Vibration in High-Speed Trains

The aerodynamic force in tunnel sections of high-speed train travel is significantly greater than in the open sections, and it is accompanied by vibrations that increase with the aerodynamic force. The vibration increases from the trains leading edge to the tail, and the yawing vibration has been found to be the dominant vibration mode as the train goes through a tunnel [8-4]. A pressure wave is initiated when the train enters a tunnel. This pressure is propagated all along the length of the train set to the tail, where it generates a large pressure fluctuation. Additionally, continuous pressure fluctuations are experienced along the length of the train as it goes through the tunnel, with the pressure on the tunnel wall side (the side nearest the tunnel wall) steadily greater than on the tunnel center side (the side nearest the center of the tunnel). The difference between the tunnel wall side pressure and the tunnel center side pressure acts on the train as an aerodynamic lateral force, which produces a yawing moment (Fig. 8.4).

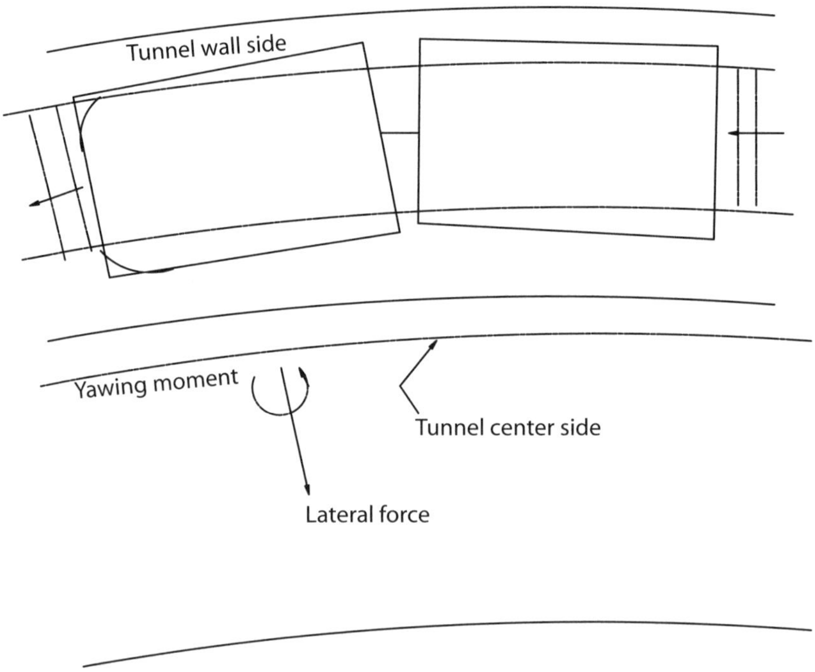

Fig. 8.4 Aerodynamic lateral force and yawing moment generation

Efforts to reduce the effects of the lateral force, hence the yaw, include introducing a yaw damper between cars to reduce the effect of the damping moment, introducing semi-active suspensions to reduce vibration by controlling damping coefficients, designing a long, streamlined nose to reduce the lateral force itself (hence the pressure difference between the wall side and the center side pressures), rounding the bottom corners of cars, and including fins in the underbody of the cars.

These same corrective measures apply to a train in an open environment but subject to cross wind. If the train is in a curved path concave to the wind, the tremendous weight (>800,000 lbs) of the train combined with the fact that the center of rotation is now in the windward side keeps the train stable, relatively unaffected by the wind. If the train is in a curved path that is convex to the wind, then the center of rotation is in the leeward side, and the train will be subject to a lateral force, a resulting yaw moment, and the accompanying vibration. The worst case is for a train in a straight path. The same wind pressure would generate a greater lateral force and a resulting greater yaw with the accompanying vibration than with the train in a convex path.

8.4 Freight Train

Freight trains are still a workhorse of the industrial age. They haul over 50 cars and containers at once. It is not uncommon to have one hundred container vessels in one haul. For a given engine horsepower, the more cars, the lower is the train speed. Sometimes two locomotives are required to haul a freight train (Fig. 8.5). Freight trains travel at about 40 mph on the average. They rarely exceed 70 mph. At such limited speed, there is minimal aerodynamic disadvantage in its typical blunt nose design. There is therefore not much aerodynamic consideration given to the locomotive designed for cargo haulage. It is, in fact, difficult to design a freight train for aerodynamics because the locomotive hauls whatever cartage is attached to it—and containers come in various shapes. Nonetheless, the small gain in an improved nose design of the freight train can translate into significant economic benefit, especially at intervals when the train travels at the high end of its speed range.

Fig. 8.5 Freight train

8.5 Aerodynamic Considerations in Engine Module Design

The significance of aerodynamic resistance as train speed increases is shown in Fig. 8.6. As with other ground vehicles, the aerodynamic drag increases with speed.

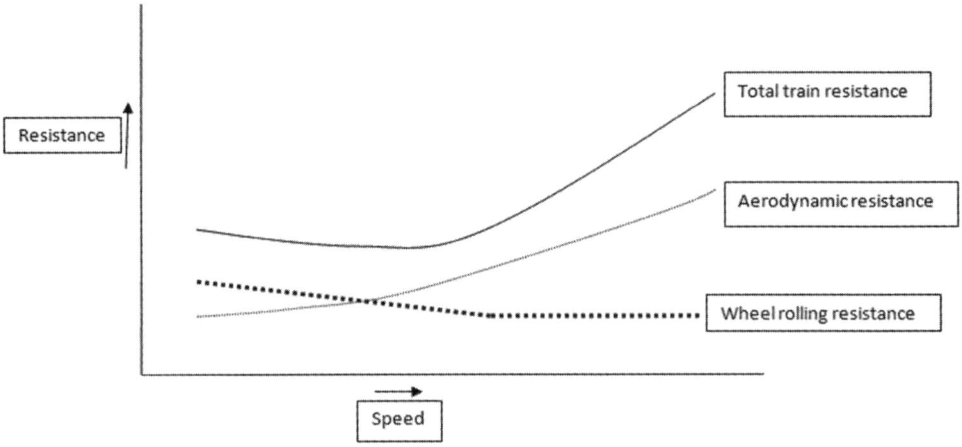

Fig. 8.6 Resistance versus speed

The rolling resistance, however, decreases slightly as the train goes faster. This is because the coefficient of friction for steel wheels to dry steel rails ($\mu_k = 0.51$, which is lower than the coefficient of friction between tire and road, $\mu_k = 0.7$) is reduced as the wheel temperature rises due to increased rail speed. It continues to decrease until it reaches a terminal value.

The aerodynamic power, consistent with the aerodynamic drag, increases steadily and becomes the dominant determinant (over mechanical power) of power required for the train (Fig. 8.7).

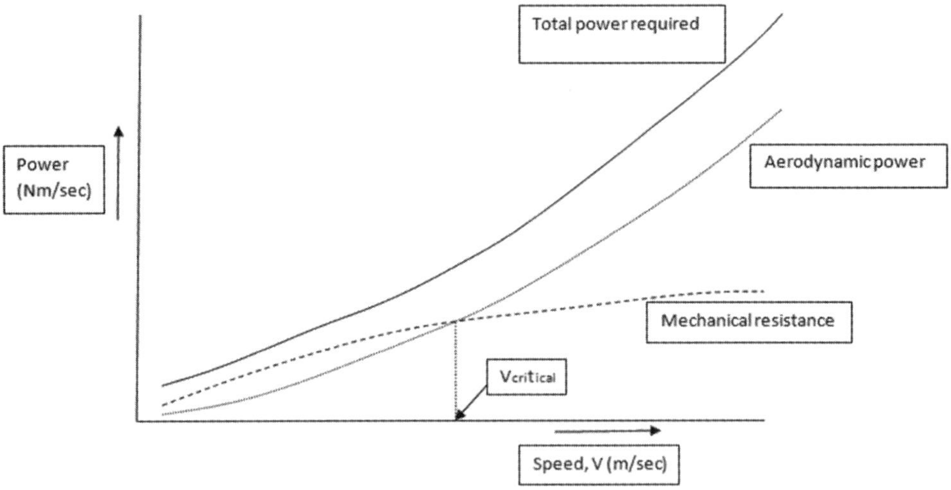

Fig. 8.7 Power versus speed

The engine module of a high-speed train is alone responsible for over 40% of the train aerodynamic drag. It is therefore essential to pay meticulous attention to its design, perhaps more than the design of the cars, if there ever will be any reason to choose between both options. A long, streamlined nose has proven to be the best design for the front of the engine module.

8.6 Aerodynamic Considerations in Train Car Design

In comparison with a freight train, a passenger train is short, and the coaches are uniform, compact, and streamlined. Also, the intercoach air gap is small—less than the critical value

$$G_{critical}/\sqrt{A} = 0.5$$

where G is the intercoach gap, and A is the coach projected frontal area.

Modifications to improve the aerodynamic efficacy of high-speed train cars include rounding the bottom corners of cars, and adding fins to the car underbody.

As in multitrailer trucks and in accordion buses, the specific aerodynamic force of a multicar train in a straight path decreases with the number of cars. The increase in the drag coefficient is mostly due to the skin friction drag of the added cars.

Summary

Trains are primarily classified as passenger trains and cargo (freight) trains. Passenger trains are further classified into citywide, regional, and intercity commuter trains. Citywide and regional service trains are low speed (less than 70 mph). Intercity trains are generally high speed, with speed as high as 200 mph in some parts of the world. Cargo trains are poor in aerodynamic design but do not suffer the aerodynamic disadvantage of such design because they travel at low speeds (usually under 50 mph).

The aerodynamic drag of a high-speed train is significantly reduced when the nose of the engine is streamlined. The aerodynamic drag increases with speed. At a critical value of about 65 mph, the aerodynamic drag starts to surpass the train's mechanical drag.

The aerodynamic power, consistent with the aerodynamic drag, increases steadily and becomes the dominant determinant (over mechanical power) of power required for the train

Lateral force and yaw moment are incurred when a high-speed train passes through a tunnel because of the pressure difference resulting from the higher pressure on the tunnel wall side over the tunnel center side. The longer the tunnel, the greater is the side force and the yaw moment.

Passenger cars should be arranged so that the intercoach gap is less than the critical value

$$G_{critical}/\sqrt{A} = 0.5$$

where G is the intercoach gap, and A is the coach projected frontal area.

Key Terms

Freight train: A train used exclusively for the transport of cargo.

High-speed train: A train designed to travel at speeds higher than 80 mph; some high-speed trains travel at speeds over 200 mph.

Low-speed train: A train designed to travel at speeds below 60 mph; a low-speed train could be a passenger train or a freight train.

The other-train: The passing train or another train in the area of influence of the subject train; the area of influence is sometimes greater than the boundary layer.

Passenger train: A train used exclusively for the transport of passengers.

Slipstream: A boundary layer along the path of an object's travel.

Subject train: The train that is of primary interest.

Review Questions

1. Classify trains by their functions.
2. Classify trains by their operational speed.
3. Is aerodynamic design crucial for a cargo train? Explain.
4. High speed is generally desirable. Mention and describe two aerodynamic features that are likely to discourage a passenger from boarding a high-speed train.

Application Questions

1. Apart from streamlining the nose of the train, what other modifications to the engine module do you think will help to reduce the aerodynamic drag? (You may apply lessons learned from race car fascia design.)

2. It is recommended that the intercoach gap be less than the critical value. Suppose you are constrained to design a train for multiple missions, and it turns out that the intercoach gap has to be higher than the critical value. How do you compensate for the aerodynamic drag penalty that this violation incurs?

Exercises

1. A train consists of 50 railcars. If each railcar is 50 ft long, 9 ft 6 in. wide, and 11 ft 9 in. high, find the total skin friction drag on the cars. What should be the maximum gap between the cars for desirable aerodynamic performance?
2. The wind is blowing at 40 mph in a head-on direction that is 35° from the normal to the direction of a high-speed train. What is the side force on a 50 ft long by 12 ft high by 9 ft wide car in the train set? What is the skin friction drag on the car?

Projects

1. Arrange a seven-car model train on a straight track. Measure and record the force required to move the train at 20 fps (or a speed of your choice). Start to remove one car at a time from the train. Each time you remove a car, run the train at the same speed that you started with (20 fps or your earlier selected speed), and measure the force required to maintain that speed. Plot the force against the number of cars. Discuss your result.
2. Repeat the preceding experiment for (a) a headwind, (b) a tail wind, and (c) wind in a direction perpendicular to the train path. What do you observe about the effect of the wind on the required force for the train operation? Does the effect of the wind exacerbate or intensify? Is there a relationship between the with increasing cars of the train?
3. Do the preceding experiment for three different types of engine modules, namely a high-speed engine module, a citywide engine module, and a freight engine module.

References

1. McCallen, R., Browand, F., and Ross, J. (eds.), *The Aerodynamics of Heavy Vehicles: Trucks, Buses, and Trains,* Springer-Verlag, 2004.
2. Paul, J. C., Johnson, R. W., and Yates, R. G., "Applications of CFD to Rail Car and Locomotive," www.airflowsciences.com.
3. Raghunathan, R. S., Kim, H.–D., and Setoguchi, T., "Aerodynamics of High-Speed Railway Train," Progress in Aerospace Sciences, Paper 38, 2002, pp. 469–514.
4. 8-4. Suzuki, M., "Flow Induced Vibration of High-speed Trains in Tunnels," the Aerodynamics of Heavy Vehicles: Trucks, Buses and Trains, United Engineering Conference, Monterey, California, pp. 443–452, Dec. 2–6, 2002.

Chapter 9
Severe Service and Off-Road Vehicles

Severe service and off-road vehicles are traditionally designed for function without consideration for the aerodynamics of the vehicle. The probable reasons for this near-nil aerodynamic consideration are the fact that severe service and off-road vehicles typically operate at low speeds—below the speeds at which the benefits of aerodynamic improvements are found to be significant, and most severe service and off-road vehicles have considerable downtime. Nonetheless, there are aerodynamic benefits, albeit moderate, that can be realized by strategic modification of certain parts of the vehicles. For example, the underbody of garbage trucks can be redesigned for drag reduction even at their relatively low operating speeds.

Severe service vehicles include (but are not limited to) construction and hauling trucks, garbage collection trucks, farm machinery, and trucks used in utility services. Off-road trucks spend very brief (transit) time on the highway or on the main roads. They include carts used within large manufacturing plants, farm machinery, and some construction vehicles. As seen in the preceding examples, a vehicle can be classified both as *off-road* and as *severe service*.

9.1 Construction Trucks

Trucks commonly used in the construction industry are rugged and primarily utility-based. Aerodynamic considerations would be viewed merely as aesthetics and hence irrelevant for their use (Fig. 9.1). Nonetheless, there are opportunities for aerodynamic improvement in the front and on the sides.

Fig. 9.1 Construction truck

9.1.1 Front Design

The front of most construction trucks is a bluff body (Fig. 9.2). The aerodynamic profiling of the front section will reduce the drag coefficient by up to 0.08%. Although the front edges are rounded, the radius is too small to prevent flow separation at the fascia. A flat face against the flow incurs drag. The front of the truck needs to be rounded to reduce the drag. Instead of a sharp leading-edge bottom, a smooth transition from the leading edge into the bottom of the cabin—just before the front axle—will reduce the truck underbody drag [9-1].

Fig. 9.2 Construction truck front

9.1.2 Side Design

Unlike in passenger transport vehicles, where most of the drag comes from the front section, the drag contributions by the side of a severe service vehicle can be nearly as much as that from the leading edge, depending on the attachments on the vehicle's sides. Apart from the attachments, the loaders are usually rectangular, with widths greater than the frontal width of the cabin (Fig. 9.3). This creates a flat face against the air flow from the sides of the cabin. Some of the flow is reflected back to the cabin sides, some dispersed laterally, and a small portion is pushed backward by the free stream air.

Fig. 9.3 Construction truck: side view

9.2 Garbage Collection Trucks

Except for the long haulers, garbage collection trucks typically make frequent stops. There is therefore not enough time for initiated flow to develop a sustaining profile. Also, like construction trucks (perhaps even more than in construction trucks), there are attachments all around the main vehicle. Most garbage trucks are rear loaders (Fig. 9.4), but there are front loaders, side loaders, and top loaders. Considering the paraphernalia all around garbage trucks, and the fact that they travel below 30 mph in over 85% of their service time, it is unlikely that aerodynamic gains will be made in modifying the cabin or loader of a garbage truck. However, some gains can be made in the improvement of the underbody. A smooth transition of an installed underbody panel, from the leading edge (behind the bumper) to the longitudinal position of the front axle, will reduce the underbody drag (see Section 6.4).

Fig. 9.4 Garbage truck

9.3 Farm Machinery

Farm machinery comes in various designs (Fig. 9.5). The design of farm machinery is almost exclusively functional. It is perhaps in the tractor design that one can expect an aerodynamic gain. The near-flat front of the engine cowling contributes to drag. So does the "square glass box" of the cabin. A pull-back of the windshield top would enhance a relatively smooth transition of the flow from the hood to the top of the cabin (see Section 6.1.2). The travel time of the flow on the top of the cabin is minimal. There is therefore no notable aerodynamic benefit in aerodynamically profiling the top of the cabin.

Soiling is ever present around farm machinery. Any aerodynamic benefits obtainable by modifying the sides of a tractor are nullified by the drag incurred by soiling. There is therefore no practical benefit in modifying the sides.

Fig. 9.5 Farm machinery

9.4 In-Plant Carts

The common in-plant carts are designed like jumbo match boxes, with slight rounding at the leading-edge corners (Fig. 9.6). The flat front contributes significantly to drag even at in-plant speeds, which can be up to 20 mph. A rounding of the front improves the flow of air at the leading edge, hence the reduction in drag [9-2]. The significance of such improvement is in the downtime of a battery-run cart and in the reduction of in-plant air pollution of a fuel-powered cart.

Fig. 9.6 In-plant cart

9.5 Sections of Severe Service and Off-Road Vehicles

Many vehicles that function as severe service vehicles are not designed from scratch as exclusively for severe service. What makes them functional as severe service is not always obvious because sometimes it is an enclosed mechanical or electrical device or a more durable component with the same geometric design specifications as the regular-use component. It is therefore to be expected that the main body profile of a severe service vehicle is not necessarily different from that of a vehicle for nonsevere service function.

The sections are essentially the engine and the attachment. The attachment can be a loader, a cart, a trailer, a farm tool, or any other equipment used in performing the severe service and off-road task. The engine drives the attachment and is housed in the cabin assembly. It could be in the hood in front of the cabin, under the cabin, or beside the cabin.

It is difficult to separate some off-road vehicles into two longitudinal parts. An armored tank, an in-plant cart or tugger, and (arguably) a four-wheeled, all-terrain sport vehicle are examples of off-road vehicles that *come in one piece*.

9.6 Sectional Aerodynamic Design

As pointed out earlier, most of the opportunities for aerodynamic improvement in severe service and off-road vehicles are found in the *engine* section. Little can be done with the *attachment* section because of the various functional tools added to this section of the vehicle.

If the engine is in front of the cabin, three opportunities are available for aerodynamic improvement: the hood, the cabin, and the front underbody. Rounding leading edge and sides of the hood and raising the hood on a slow slant from the nose to the rear at the hood-windshield interface will result in the reduction of drag. Rather than having a windshield that stands almost perpendicular to the hood, a slanting windshield from the hood-windshield interface to the roof of the cabin improves the aerodynamics of the flow on the cabin. For engines placed under the driver's seat, the hood no longer exists as a distinct entity. Rather, it is considered as part—the lower part or rear part, depending on the location—of the cabin. Aerodynamic improvements then focus on the cabin and the underbody.

9.7 Underbody Design

The underbody of the cabin is generally small in length compared with the length of the attachment. Yet, because of its location, it influences the rest of the underbody flow beyond its boundaries. Therefore, it is important to design the nose of the leading edge for a smooth transition of the downward flow to the underbody.

9.8 Soiling in Severe Service Vehicles

In most severe service vehicles and perhaps in all off-road vehicles, soiling is accepted as inevitable (Fig. 9.7). However, a reduction in soiling is desirable in severe service vehicles, though for operator comfort rather than for vehicle drag reduction. For this reason, the efforts in the reduction of soiling are focused on the front section (usually the cabin) of severe service vehicles.

Soiling adds to vehicle drag, and because soiling builds up over time, so does the vehicle drag. A practical approach to the reduction of soiling is to wash the vehicle at frequent intervals.

Fig. 9.7 Soiling

Summary

Severe service and off-road vehicles are traditionally designed primarily for function. Little attention is paid to the aesthetics or aerodynamics of such vehicles. Most severe service and off-road vehicles operate at low speeds—below the speeds at which the benefits of aerodynamic improvements can be significant. They also have a considerable downtime. Examples of severe service vehicles are construction and hauling trucks, garbage collection trucks, farm machinery, and trucks used in utility services. Off-road trucks spend very brief (transit) time on the highway or on the main roads. They include carts used within large manufacturing plants, farm machinery, and some construction vehicles. Some vehicles function as off-road and severe service vehicles.

From a design standpoint, severe service vehicles are essentially two-component vehicles—the engine-cabin component and the attachment component. Opportunities for aerodynamic improvement are more likely to be found in the engine-cabin component. The hood, the windshield, and the underbody of the engine-cabin module are areas that can be modified to enhance the aerodynamics of severe service vehicles.

Almost all off-road vehicles—more so than severe service vehicles—function in a continuous soiling environment. The extent to which soiling is addressed in severe service vehicles is primarily for passenger comfort. Soiling adds to vehicle drag, and because soiling builds up over time, so does the vehicle drag. A practical approach to the reduction of soiling is to wash the vehicle at frequent intervals.

Key Terms

In-plant car/cart: A motorized cart driven within (usually) a large manufacturing plant; it is used to transport personnel and small components from one station within the plant to another.

Off-road vehicle: An off-road vehicle is a vehicle that spends a very small (or zero) percent of its operational time away from the main roads and highways.

Severe service vehicle: A severe service vehicle is one that is used more frequently, and/or employed in more rugged use than the average vehicle in the same engine category.

Review Questions

1. Why has there been no serious aerodynamic consideration in off-road vehicle design?
2. How is an off-road vehicle different from a severe service vehicle? How are they similar?
3. What is the best approach to ameliorate the effect of soiling in off-road and severe service vehicles?

Application Question

It is important to design the nose of the leading edge for a smooth transition of the downward flow to the underbody. Why?

Exercises

1. What are the design challenges in breaking a severe service vehicle into sections as is readily done in cars? Into how many longitudinal sections can one reasonably group the components of a severe service vehicle? Name them.
2. Severe service and off-road vehicles suffer the same cross-wind obstruction as cars. How would you modify a severe service vehicle to reduce the effect of cross wind?

Projects

1. Modify the front section of a model severe service vehicle for aerodynamic improvement. Determine the drag reduction. Then modify (a) only the fascia, (b) only the hood, and (c) only the front underside. Find the drag reduction due to each modification. Sum up the respective drag reductions and compare these with the drag reduction from the collective modification of all three components. How large is the difference? Explain the difference. Which component appears to be most significant for aerodynamic drag reduction?
2. How would you design the underbody of a severe service vehicle so that soiling is easily detached therefrom, thus reducing the dead weight that such vehicles carry in the winter and in the rainy season?

References

1. Mason, Jr., W. T., and Beebe, P. S., "The Drag Related Flow Field Characteristics of Trucks and Buses," *Aerodynamic Drag Mechanisms of Bluff Bodies and Road Vehicles,* Sovran, G., Morel, T., and Mason, Jr., eds., Plenum Press, 1978.
2. Carr, G. W., "Potential for Aerodynamic Drag Reduction in Car Design," Dorgham, M. A., ed., *Impact of Aerodynamics on Vehicle Design, Technological Advances in Vehicle Design Series,* SP3, Interscience Enterprises Ltd., 1983.

Chapter 10

Race Cars, Sports Cars, and Convertibles

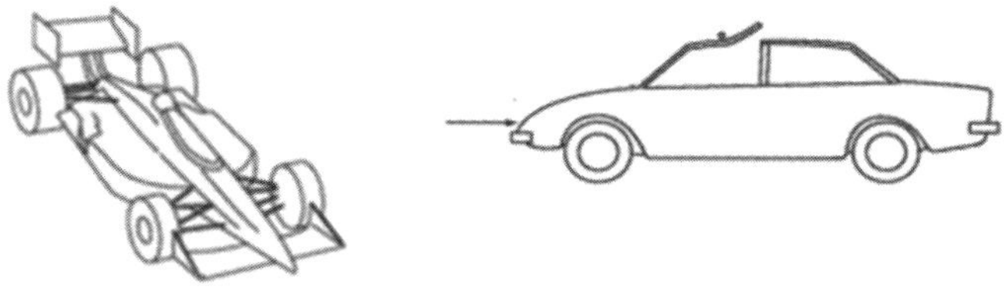

In the quest for range and speed, modern cars continue to evolve, with a focus design or an optimum design. Optimum design cars are built for a wide range of road conditions and services at reasonable, operational speeds. Most sedans and station wagons are essentially optimum design cars with emphasis on service, range, endurance, durability, and less emphasis on speed. Because of the growing cultural interest in obtaining speed, cars designed for emphasis on speed have grown from just a few in the 1900s to now over five classifications.

Typically, aerodynamic forces are the main factors affecting the control of race cars. Brake handling, driver control, inertia forces, and movement of the center of pressure (c.p.) relative to the center of gravity (c.g.) are examples of the concerns that affect the stability of a car, and these are mostly sensitive in race cars and in motorcycles. Convertibles are drag-embracers by design, with a significant amount of the drag increase coming from the roof section (or the absence of roof section) design. The open-air concept in convertibles and topless cars is a luxury that is paid for in escalated aerodynamic drag of the vehicle. A sports car is typical a small, light-weight, two-door, two-seater car designed for high performance beyond that of a sedan, approaching that of a race car. Sports cars combine high performance (a design feature of race cars) with passenger comfort (a design feature of sedans). Sports cars are often used interchangeably, sometimes without modifications, as race cars and for everyday utility cars.

10.1 Race Cars and Sports Cars

Perhaps the most important aerodynamic goal in a race car or sports car design is the maximization of the downforce. At high speeds, race cars generate a high downforce.

$$\{\text{downforce} \propto (vehicle\_speed)^2\}$$

The high downforce increases the normal force on the tires, and this in turn increases the car's braking performance. The significance of tires in race car performance is captured by the commonly expressed belief that race car design is all about optimizing the use of tires [10-1]. The significance of the downforce is underlined when we consider that for a race car at an initial speed of about 250 km/h, a downforce of $C_L = -2$ results in a reduced braking distance of about 30 m, from a breaking distance of 120 m for zero downforce—a 25% reduction [10-2]. The turning speed is also enhanced by the downforce. The higher the downforce that a race car can generate, the greater is the speed at which it can safely negotiate a turn.

The underbody of a race car is shaped with channels (air dams) to streamline air flow and to increase the underbody effective surface area, both of which result in increased downforce and improved engine cooling (Fig. 10.1).

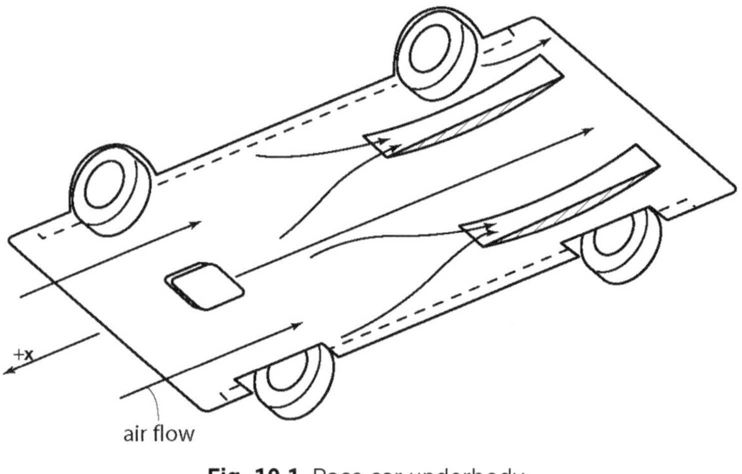

Fig. 10.1 Race car underbody

Externally mounted inverted aeronautical mini-wings are common negative lift devices incorporated into the design of race cars. Their intended function is to increase the downforce. The vehicle wheels are pressed more to the ground due to the higher pressure on the top of the mounted wings. This improves traction and cornering characteristics of the vehicle. For practical reasons, the wings are typically mounted on struts attached to the car's main frame. Wings placed in the front are referred to as front deck mounts and those at the back, rear deck mounts (Fig. 10.2). The front deck mounts primarily help to streamline the starting flow, whereas the rear deck mounts reduce the lift on the rear wheels.

Fig. 10.2 Front and rear inverted wings

The positioning of the rear deck mount is very critical. If positioned too far to the rear, as shown by the dashed line airfoil in Fig. 10.3, it certainly increases the traction of the rear wheels but it would also create a turning effect about the point of contact of the rear wheel to the ground, resulting in a nose-up tendency and a consequent loss of control. Similarly, low positioning produces a comparatively ineffective device and so loses its raison d'etre. In addition, the wake behind it, when merged with the wake behind the race car, causes instability and poor driver control [10-1].

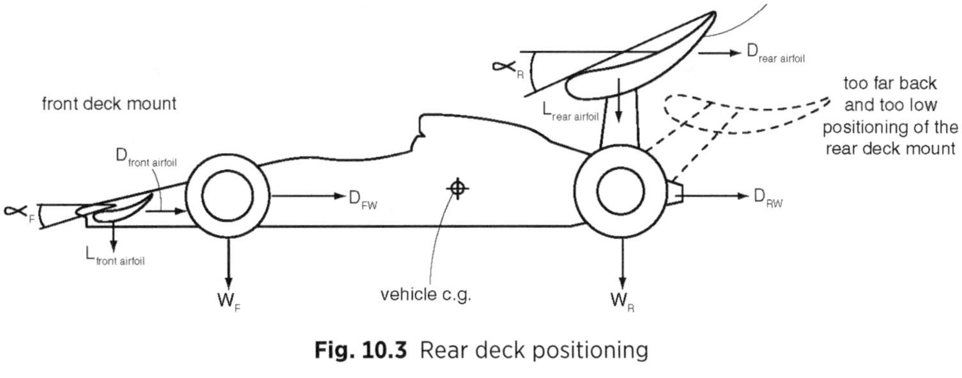

Fig. 10.3 Rear deck positioning

Spoilers, common in sports cars and in race cars, can be considered as functional rather than geometric mini airfoils. A spoiler is often made of just sheet metal without shaping it into an airfoil. Functionally, the front spoiler or front underbody dam (Fig. 6. 13(a)) laterally diverts the air flow ahead of it, thereby reducing the underbody drag and the pressure under the hood. The reduction of the underhood pressure has the dual benefit of enhancing the performance of the cooling system and reducing the lift at the car front. In stock cars, a leading-edge flat plate is placed ahead of the front spoiler. It serves to redistribute the flow at the leading edge such that the longitudinal component is increased and the lateral distribution is reduced (Fig. 10.4). The drag at the fascia is also reduced as the air flow over the car is more streamlined due to the presence of the plate. The plate must be very low at the leading edge (Fig. 10.5). Otherwise it would generate lift, which is highly undesirable.

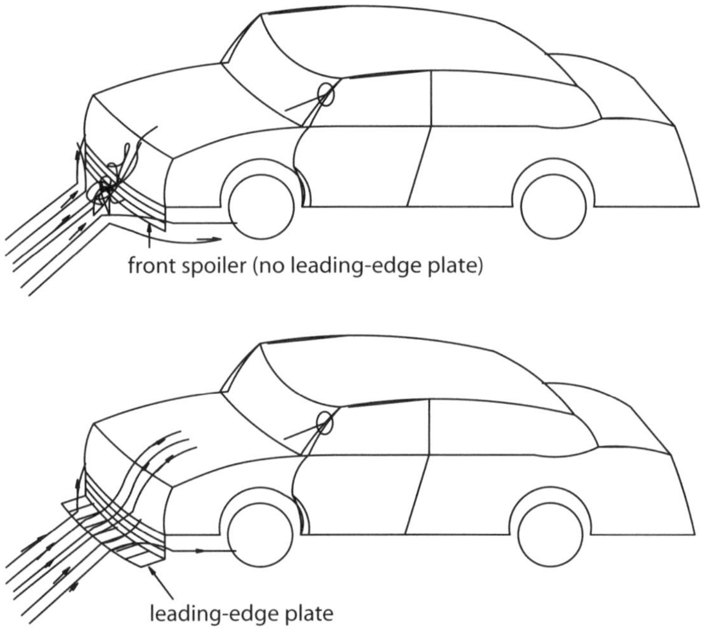

Fig. 10.4 Flow distribution with and without a leading-edge flat plate on the front spoiler

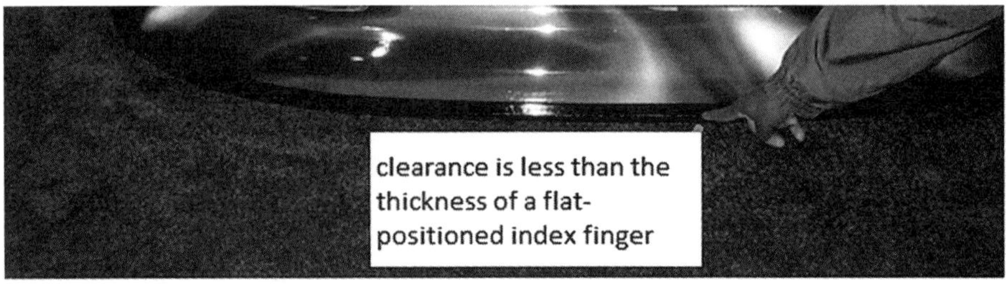

Fig. 10.5 Leading-edge flat plate

In a Formula One car, the front wing is the leading-edge plate. It is, however, a more sophisticated design. The front wing of a Formula One car is a well-tested, carefully selected inverted wing formed from a moderately high-speed subsonic airfoil section (Fig. 10.2).

As the wing gets closer to the ground, it generates more downforce and greater drag. The downforce and the drag both reach respective critical values, after which they start on the diminishing trend, with the downforce showing a greater sensitivity as the wing continues to get closer to the ground [10-3]. The rule of thumb is to have the ground clearance be 10% of the wing chord length.

As shown earlier in Fig. 6.11, the rear spoiler is placed at the back, attached to the top of the trunk. It serves to increase the downforce of the vehicle. Rear spoilers usually range in height from 2 in. to about 4 in. and are at an inclined angle of between 120° and 150° to the trunk tangent line.

Endplates are typically added at the rear to the ends of a rear airfoil (Fig. 10.2). They serve to improve the downforce near the wing tip by preventing downwash. Although not the primary reason for their inclusion, they increase the aspect ratio and hence the downforce.

AR_p, the aspect ratio with endplates is given by

$$AR_p = AR(1 + \kappa h/b)$$

where

AR is the aspect ratio of the rear airfoil before the addition of endplates;
h is the height of the endplates;
b is the wing span; and
κ is a correction factor.

Side force, due to a sudden gust or a sustained side wind, is one of the obstacles to the performance of a race car. To overcome side forces, it is important to design the car such that its c.p. trails the c.g. on the longitudinal axis of the vehicle. The larger the static margin, the greater is the stability of the vehicle.

In a race, race cars travel in a pack at very high speeds. The driver behind the lead car can experience a sudden pull of the vehicle because it is saved from the drag experienced by the leading car. Race car drivers take advantage of being in the slipstream of the leading vehicle to accelerate their cars at no fuel consumption cost [10-2, 10-4, 10-5].

10.1.1 Race Car Design Envelope

Race cars are designed for speed. The design therefore focuses on devices to increase the speed and improve (or at least maintain) the handling and stability as the speed increases. Front wings, rear wings, splitters, shark fins, roof strips, recessed side windows, underbody side skirts, spoilers, and dams are a few of the many devices incorporated in race cars to reduce the drag and to increase the stability.

Race cars have between-service life requiring parts replacement almost after every race, and an overhaul every few months.

10.1.2 Performance and Dynamic Stability

Vehicle performance is highly improved by increasing the downforce on the tires. The tires of a fast car must grip the road firmly so that they do not slip when the car accelerates or when the driver undertakes some maneuvers. Sometimes, the back tires are wider than the front tires for improved track grip. The narrower front tires help to reduce the drag—the wider the tire, the greater is the drag. Smooth tires are used on dry surfaces but treaded tires are best for wet surfaces, where slippage can readily occur without the stronger grip that the tread provides. Fully covered wheels are preferred to open wheels because they improve drag on the wheel assembly when compared with open wheels or with partially open wheels with radial spokes.

10.1.3 Control Surfaces

Inverted wings serve to increase the downforce. They are perhaps the most vivid of the control surfaces of a race car. Their low placement at the leading edge reduces the left and increase the downforce. At ground clearance below one-tenth of the wing chord length, the downforce is reduced due to the clearance taking on the form of a venturi. Therefore the leading edge wing is placed low only up to a critical point so that the ground clearance does not result in lift generation (Fig. 10.6).

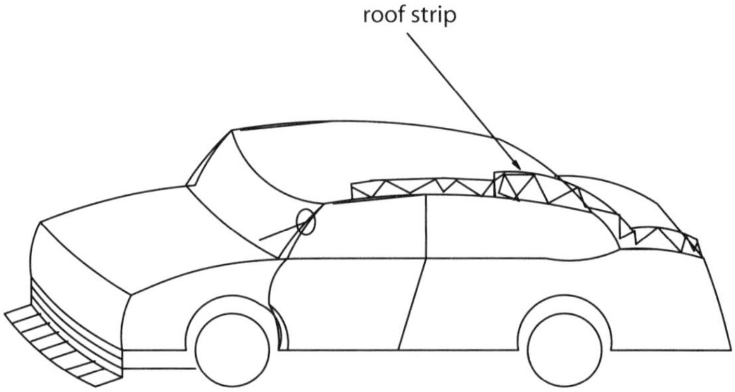

Fig. 10.6 Control surfaces

10.1.4 Soiling in Race Cars

Race cars travel over specially designed roads. The roads are dry, very well graded, and regularly maintained so that there are no bumps or pits that can, in the least, disrupt the stability of the vehicle and the streamlined flow around the vehicle or, much worse, cause vehicle and human loss. Because of the meticulous design of the road and the control of the racing environment, soiling is not as much of a concern as with general purpose vehicles. Clarity of the windshield is perhaps the first concern when considering soiling. Some race cars do not have any windshield on the vehicle.

10.1.5 Race Car Overall Aerodynamics

The overall aerodynamics goals in race car design are to increase the downforce on the vehicle, thereby increasing the dynamic stability and improving the performance, and to reduce the drag on the vehicle. Improvements for drag reduction are mostly performed on the front section, and stability devices are more prevalent in the rear section. The trend is to make the whole car become a downforce generator.

Many stock cars no longer have doors. The absence of doors helps to improve the streamline flow on the vehicle sides, thereby reducing the drag. Decals are drag inflictors, but they are of a business essence. Therefore, decals are applied at about the same level as paint, and a clear coat is applied to the entire surface. Large decal wraps are sometimes used as an alternative to paint.

10.1.6 Supercars

Built for style and speed, supercars are fast, powerful, and expensive sports cars—perhaps the fastest, most powerful, and most expensive cars on the road. Much effort is put into the cooling system and the body aerodynamics of supercars. Among the well-known supercars are the Jaguar XJ 220, the Ferrari F 50, the McLaren F1, Lamborghini, and Porsche GT RS. The engine is often rear-mounted so as to increase the car's dynamic stability. The weight of the engine pushes the back wheels down on the road, thereby increasing the road-grip of the tires. Supercars generally can also be used for everyday driving and for occasional racing.

10.2 Rally Cars and Dragsters

Rally cars, like SUVs, are road vehicles modified for all-weather, all-terrain endurance.

Dragsters are blue-smoke-spewing cars with humongous engines. Dragsters, like supercars, are designed for speed, but unlike supercars and race cars there is little aerodynamic emphasis on the engine compartment design. The engines of dragsters typically protrude quite distinctly from the body. Although this design is perhaps the only safe approach to meet the smoke-spewing requirement, it nonetheless adds tremendously to the vehicle drag.

10.3 Convertibles

Convertibles are inherently drag embracers. Although the drag contribution due to the roof is eliminated, the effect of the drag on the sides is greater than if there had been a rooftop. The reversed flow at the top of the wind screen and on the sides creates pockets of turbulence at various spots in the passenger compartment. Additionally, when the flow impinges on a structure or a person in the car, it impacts with a direct force, F_i, which acts opposite to the vehicle heading (Fig. 10.7). A sunroof is an aerodynamic compromise between a hard top and a convertible.

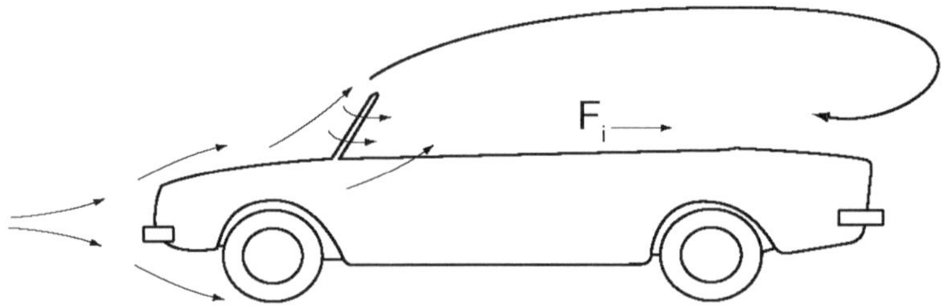

Fig.10.7 Convertible

10.3.1 Side Windows

The side windows in a convertible are themselves noise generators and drag enhancers. They do not have the car roof to provide a flow boundary to the upward moving air component along the glass, to seal the inside from the top edge spillage of outside air flow, and to provide a structural support that reduces the deflection of the top portion of the glass in cross wind.

10.3.2 Sunroof

A sunroof can be installed either as a deflector roof (Fig. 10.8) or as a slider (Fig. 10.9). Most modern factory-installed sunroofs are adjustable to either the deflector or the slider. In the deflector setting, the flow is suddenly raised as it passes over the glass plate. It travels unattached for a portion of the distance on the roof. It then lands at a longitudinal position near the back seat or in some cases as far back as aft of the trunk. The greater the tilt angle to the rooftop, the longer it is for the flow to reattach (or to descend). At the top edge of the sunroof, there is a flow reversal that causes a lot of noise as it hits the sunroof. The perturbation is higher as the vehicle speed increases.

Fig. 10.8 Deflector sunroof

The slider sunroof does not interfere with the flow pattern at the rooftop as much as is the case with the deflector sunroof. The flow outside the wall region is unaffected. But at the wall where there is now no surface when the sunroof slides open, there is a little depression. The airstream from the depression component is deflected into the car and continues rearward in the car. It hits the rear windshield and is reflected to the front, hitting occupants in the front section of the vehicle.

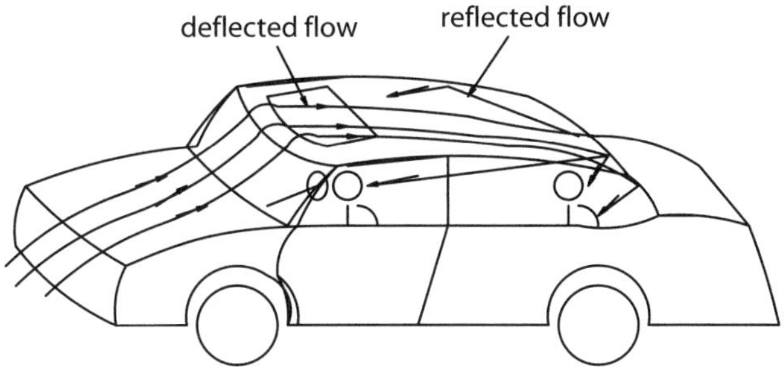

Fig. 10.9 Slider sunroof

10.3.3 Cloth-Top

Cloth-tops may be fanciful, but they are aerodynamically the worst roof design. In addition to the tremendous drag due to the cloth material (except if the material is silk) the absence of rigidity (or even of firmness) causes a lot of flutter and noise. Cloth-tops are generally not intended for use in vehicle operation. Rather, they are mainly for vehicle protection from the elements such as rain or snow, and this is why cloth-tops are associated with convertibles.

Summary

Aerodynamic forces are very important in the control of race cars. Brake handling, driver control, inertia forces, and relative movement of the center of pressure to the center of gravity are among the concerns in addressing the stability of a car and are particularly sensitive factors in race cars. The most important aerodynamic goal in a race car or sports car design is the maximization of the downforce. The downforce is directly proportional to the square of the vehicle speed:

$$\{\text{downforce} \propto (\textit{vehicle\_speed})^2\}$$

Spoilers, air dams, leading-edge flat plates, front wings, and rear wings are examples of devices introduced in race cars to increase the dynamic stability and improve the performance of race cars.

Sunroofs are either deflector roofs or sliders. Deflector roofs are typically noisier than sliders.

Key Terms

Deflector sun roof: A type of sunroof design that opens up at an angle (usually below 20°) to the roof of the vehicle.

Downforce: A negative lift force; downforce aerodynamically "presses" the vehicle to the ground, thereby ensuring vehicle dynamic stability.

Front wing: An airfoil-shaped winglet at the leading edge of a race car (typically in Formula 1 and Indy race cars).

Leading-edge plate: A plate installed in stock racing cars, usually in front of the front spoiler; a leading-edge plate serves as the "front wing" for a stock car.

Rear wing: An airfoil-shaped winglet at the trailing edge of a race car (typically in Formula 1 and Indy race cars).

Slider sunroof: A type of sunroof design—a cut-out within the continuous profile of the roof.

Supercar: A very high end sports car; supercars have big engines pulling very high horsepower.

Review Questions

1. What are the differences between a sedan and a sports car?
2. Is a sports car different from a race car? Explain.
3. Name three important factors to take into account when designing a race car.
4. What is a leading edge plate? In what type of vehicle is it typically installed? How does it work?
5. Which is preferred—a deflector sunroof or a slider sunroof? Explain.

Application Questions

1. How do you think the thickness of the leading-edge plate in a stock car affects the profile of the flow?
2. Side skirts are among the control surfaces to improve the stability of a race car. Which do you think is more important if you have to choose between side skirts and a front spoiler? Explain.

Exercises

1. The rear airfoil of a race car has a span of 55 in. If a 12-in. high endplate is attached to each end, what is the percentage increase in the aspect ratio due to the addition of the endplates? The airfoil chord length is 9 in. Use a correction factor of 0.85.
2. A race car moving at 110 mph has a 100 lb downforce contribution due to its front wing. What is the percent increase in the downforce contribution of the front wing when the car reaches a speed of 220 mph?

Project

Install a front wing and a rear wing in a Formula One type race car model. Keep the location of the rear wing fixed. Change the vertical position of the front wing starting from a ground clearance equal to twice the chord length of the front wing and going as far down as possible without touching the floor. Proceed at intervals of no more than 0.1c, where c is the chord length of the front wing. Record the lift and the drag at each setting. Plot a graph of the lift versus the ground clearance. Do the same for the drag. Which parameter do your results indicate as being more sensitive to the change in ground clearance? Why do you think this is the case?

References

1. Milliken, W. F., and Milliken, D. L., *Race Car Vehicle Dynamics*, SAE International, 1995.
2. Katz, J., *Race Car Aerodynamics*, Robert Bentley, Inc., 1995.
3. Mokhtar, W. A., and Lane, J., "Racecar Front Wing Aerodynamics," *SAE International Journal of Passenger Cars–Mechanical Systems*, Vol. 1, Issue 1, SAE Paper 2008-01-2988, 2008.
4. Scibor-Rylski, A. J., "Proceeding, First Symposium in Road Vehicle Aerodynamics," The City University, London, 1969.
5. Romberg, G. F., Chianese, F., and Lajoie, R. G., "Aerodynamics of Cars in Drafting and Passing Situations," SAE 710213, 1971.

Chapter 11
Motorcycles

Contrary to a general perception, motorcycles are not relatively better fuel misers than cars. Although it is difficult to come up with a truly generalized equation for comparison, we can utilize such parameters as vehicle weight and occupant capacity for a normalized comparison of the fuel consumption rate between a car and a motorcycle. In the following comparisons, we will consider the vehicle specific weight, the passenger capacity, and the specific fuel consumption.

Weight comparison: The average weight of a motorcycle is 300 lb. The average gas mileage is 38 mpg. That means 1 gal of fuel will take 1 lb through 11,400 miles. For a car, the average weight is 3400 lb, and the gas mileage is 25 mpg. Therefore, 1 gal of fuel will take 1 lb fuel through 85,000 miles—seven times as efficient as a motorcycle.

Occupant capacity: A car has an equivalent weight capacity of six occupants: the driver and four passengers, plus the trunk space for carrying additional weight. At 25 mpg for a car, it means 1 gal of fuel powers the car with five occupants plus trunk capacity through 25 miles in an hour. If we consider the average passenger weight to be 180 lb, and recall that the gross weight of an unoccupied car is 3400 lb, then the total weight moved is $180 \times 6 + 3400 = 4480$ lb. Therefore we can state that 1 gal takes 4480 lb through 25 miles in an hour. A motorcycle normally takes two passengers at full capacity. We will therefore use the two-passenger equivalent for comparison. For a car, the value is 1493 lb. The car will take a two-passenger equivalent through 75 miles in an hour. Compare this with 38 miles for a motorcycle.

Specific fuel consumption: Finally, we will consider the specific fuel consumption. Average brake horse power (BHP) for a motorcycle is 33 HP; for a car, it is 150 HP. Fuel consumption for a motorcycle is 10.5 lb/h. This gives a specific fuel consumption value of 0.32 lb per HP-h.

For a car, the average fuel consumption is 2.5 gal per hr. At 6.15 lb per gallon this comes to 15.4 lb/h. With a BHP of 150 HP, the specific fuel consumption comes to 0.10 lb per HP-h, about three times better than the motorcycle.

In spite of the numerical results, the above comparisons primarily serve a qualitative, rather than quantitative, purpose to emphasize the need for improved fuel efficiency of the motorcycle. Since it is unlikely to obtain the sought efficiency from weight reduction, the aerodynamic avenue is probably the only area available for further improvement.

11.1 Motorcycle Classifications

There is no universal system for classifying all types of motorcycles. Therefore, a motorcycle could fall into more than one category of the many systems of motorcycle classification. The existing classifications are generally a description of how the motorcycles are put to use, or of the designer's intent, or a combination of both. The six main, widely used categories are cruiser, sport, touring, standard, dual-purpose, and dirt motorcycles. Mopeds and scooters are sometimes included in the groupings of motorcycles (Fig. 11.1).

Fig. 11.1 Types of motorcycle

A motorcycle is a street bike or an off-road bike or both, apart from its more specific function category. A brief (and necessarily loose) description of the six categories is presented here.

11.1.1 Cruisers

Cruisers are typically motorcycles with large displacement engines with tuned low-end torques. They are relatively easy-ride machines, requiring occasional shift to accelerate or to maintain vehicle control. Cruisers are indeed designed for cruising, with the foot pedal forward, and the handlebar high so that the rider's spine leans back slightly. Power cruisers are high-end cruisers—more horsepower, upgraded brakes and suspension, better ground clearance, premium surface finish, and some custom features.

11.1.2 Sport

Sport motorcycles are to motorcycles as sports cars are to cars—they are made for speed and performance. Emphasis is placed on braking, cornering, and rapid acceleration. They have high-performance engines and heavy-duty suspension systems. They are usually equipped with aerodynamic fairings that enclose the engines and with windshields designed to direct the flow of air around the vehicle rider, thereby reducing the drag.

11.1.3 Touring

Touring motorcycles are designed for range. They have large-displacement engines and large-capacity fuel tanks. They are designed for upright seating and are fitted with fairings and with windscreens that are sometimes tinted. Touring motorcycles are sometimes fitted with sidecars.

11.1.4 Standard (Street)

Standard (street) motorcycles are general purpose street motorcycles. They are designed for an upright riding position. The foot pegs are below the rider, and the handlebars are high enough so that the rider does not need to lean forward.

11.1.5 Dual Purpose

Dual purpose motorcycles are also called on/off-road motorcycles. They function as street machines and as dirt bikes. A dual purpose bike is essentially a dirt bike frame with mirrors, lights, signals, and instrument panel, as found on a street bike [11-1]. They are higher than the typical street motorcycle. They have a high center of gravity and tall seat height that allows for considerable suspension travel on rough roads. Adventure motorcycles are dual purpose motorcycles with bias toward paved road functionality.

11.1.6 Dirt Bikes

Dirt bikes are off-road motorcycles. They are relatively light-weight, rugged, with long suspensions and high ground clearance. Perhaps the most popular of bike sports is the dirt bike sport *motocross*. The goal in motocross is to have the bike move through terrains with maximum speed and minimum effort. A dirt bike is therefore designed to maximize traction and minimize the effectiveness of obstacles. Traction is maximized by a skillful application of rider weight transfer—longitudinally, laterally, and vertically. Because of the fast response required of the rider, riders in motocross typically stand on pegs for most of the course.

Braking is as important as speed is in motocross maneuvers. When braking through a rugged corner, the biker should be standing. Otherwise, the bike would tend to throw the rider up and pitch the rider forward. The front brake accounts for about 70% of the motorcycle braking efficiency; therefore, it must be used often in braking. It requires more rider response energy than the rear brake, however.

A motorcycle's maximum traction is obtained in the straight, upright attitude (just as it is for maximum braking efficiency). This positioning (standing and leaning forward in the standing posture) puts weight on the front tire, forcing the front tire onto the ride surface. An exception is when the rider is on a terrain with large bumps. In this case, the rider sits back so that the weight is kept on the rear tire. Surfaces that present a braking challenge to the rider include hard-packed clay with pebbles, hard-packed track with a film of mud on top, off-camber surfaces, and surfaces full of wet leaves and pine needles. The rider should avoid using front brakes on such surfaces. Front brakes should also be avoided when only a relatively small reduction in speed is required.

11.2 Design for Aerodynamics

The aerodynamic features of a motorcycle depend on the mission specification of the motorcycle. A motorcycle experiences the same aerodynamic forces to which a car is subjected, except for the gain in downforce. The downforce consideration does not apply in the case of motorcycles because there is practically no plan area for the downforce to act upon. Although not a bluff body, a motorcycle is essentially unstreamlined. Therefore, the pressure drag dominates the friction drag. Most of this drag originates in the frontal part and around the fuel tank [11-2].

An aerodynamically modified frontal area and a morphed fuel tank are essential to significantly reduce the pressure drag. A common solution to the frontal area design is to install an aerodynamic windshield on the motorcycle ahead of the handlebar such that practically all the motorcycle upper-body components, except the headlights, are inside the frontal area of the windshield. In motorcycles that have no windshield installed, a greater effort would be required to create an aerodynamic leading edge. This includes a curved handlebar mimicking the leading profile of a teardrop, a fuel tank that takes the planform of a low-speed subsonic airfoil, and some components crafted to have a teardrop cross section (Fig. 11.2).

Motorcycle manufacturers have recently made fairings a more or less standard feature, especially in sport motorcycles. Fairings, usually made of plastic, have the dual benefits of component protection and drag reduction. Some components are protected from contacting the ride surface. In addition to the resulting drag reduction, the aerodynamic construction of the fairing allows the fairing-equipped motorcycle to slide more easily, with less friction than its all-metal, unprotected counterparts [11-3]. An unintended safety benefit in motorcycles with fairings is the reduced deceleration when a motorcycle slides on its side in a collision. Such reduction allows for more reaction time by the rider.

Another aerodynamic improvement device is the front fender. It directs the inlet air backward, streamlining it and reducing the lift [11-4].

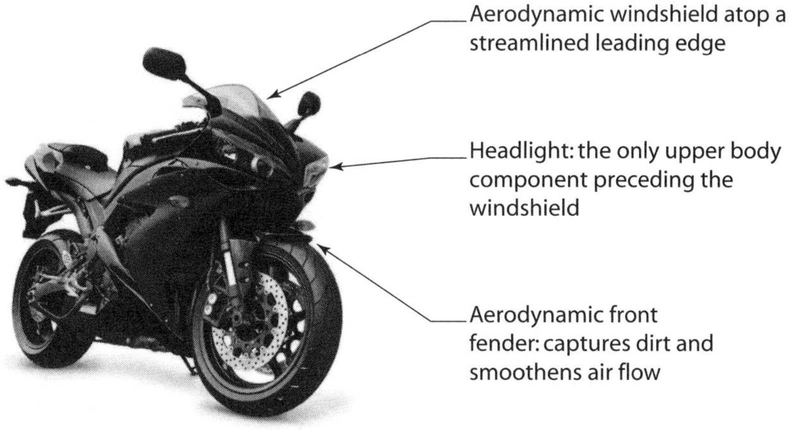

Fig. 11.2 Drag reduction design for motorcycle components

11.3 Motorcycle Stability and Control

As with race cars, aerodynamic forces have become the main factors affecting the control of motorcycles. The rider's sense of stability is generated in response to a slight change in the behaviors of the motorcycle due to disturbances or aerodynamic resistance during the ride [11-5]. The information basis for the sense of stability that the rider gets during a ride may be grouped into two main kinds:

1. The sensation—visual, auditory, tactile—that the rider gets during the ride, and
2. Past experience of the rider (such as the sense of road gripping).

Brake handling, rider control, inertia forces, and movement of the center of pressure (c.p.) relative to the center of gravity (c.g.) are particularly sensitive issues in the operation of motorcycles.

Ideally, we want low drag and high downforce (or at least, low lift), with the c.p. aft of the c.g. for stability. It is difficult to achieve a forward c.p. because most of the aerodynamic load is in the front. It is also difficult to streamline a motorcycle because of cut-outs and protruding components needed for practical reasons. The top view of an ideal motorcycle is shown in Fig. 11.3.

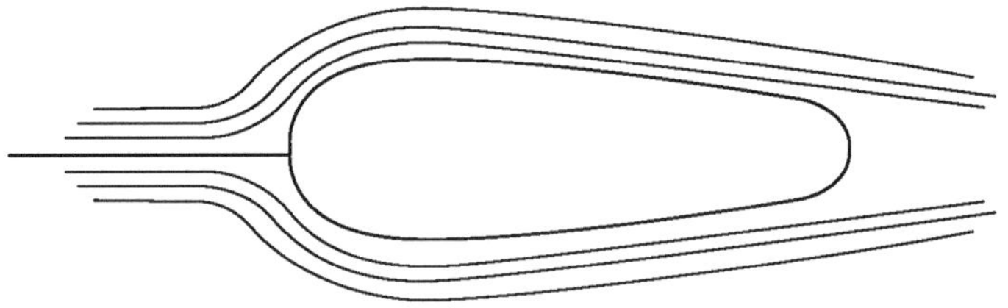

Fig. 11.3 Planform of an ideal motorcycle

The higher the speed, the greater is the fore movement of the c.p. as well as the risk of aerodynamic lift. The lift mainly comes from the pressure on the fender, the pressure on the mudguard, and the turning effect due to the drag on the front assembly (Fig. 11.4).

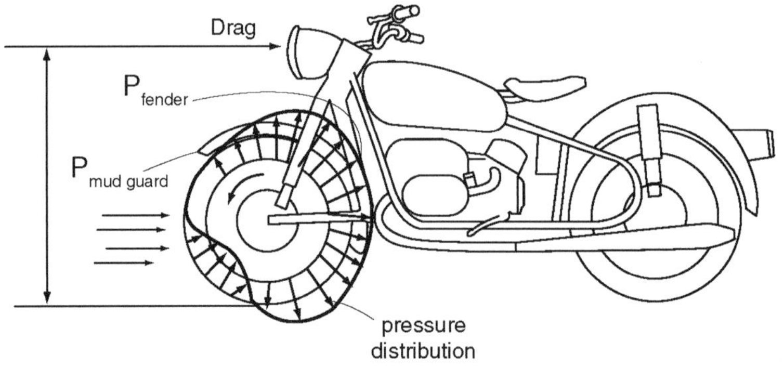

Fig. 11.4 Lifting forces on a motorcycle

The side force creates a turning effect, tilting the motorcycle laterally and away from the cross wind. The rider counters this by tilting into the wind. But in doing so, the rider tilts the weight of the rider-machine away from the centerline of the motorcycle (Fig. 11.5). The less the tilt, the more stable is the motorcycle. Therefore, the goal is to reduce the tilt necessary to counter a side force. Since there is no control over the side force that comes against the rider-machine, the design focus would be on the c.p and c.g. locations. A low c.p. reduces the turning effect of a side force, and a heavy motorcycle with a high c.g. produces an inherent balance to a strong side force.

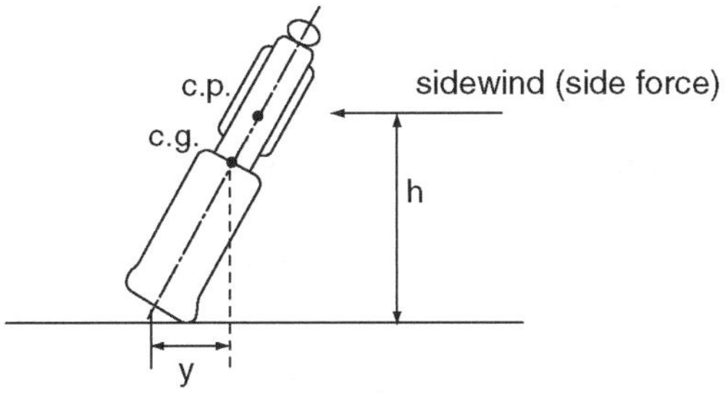

Fig. 11.5 Design for stability

In an effort to create a sleek motorcycle, designers end up creating a two-wheeled mini-car with excessive gross weight and with rider constraints that outweigh the benefits of improved aerodynamic design. The lesson in this is to seek a balance, an optimum design in which aerodynamic consideration is only one of the factors. Intangible factors such as the open air feeling, which motorcycle riders cherish, are sometimes so important that an aerodynamic sacrifice is required in the search of a balance and of the ultimate goal—market appeal. Computational methods have gained increasing acceptance as tools to optimize the vehicle performance–rider comfort combination of the motorcycle [11-6].

11.4 Soiling in Motorcycles

Unlike the occupants of a car, the motorcycle rider is not protected from the elements—wind gust, noise, and soiling. A motorcycle rider is, for all practical purposes, a driver sitting on, but detached from, the vehicle. Because the rider sits on rather than inside the vehicle, he/she is exposed to the elements just as much as the vehicle. Except for motocross, where mud, messy as it is, is part of the event, the rider would prefer a reduction in vehicle and rider soiling (Fig. 11.6). This is even more crucial than for a car, for safety reasons. A film of mud

Fig. 11.6 Soiling in motorcycles

over a hard surface acts like a lubricant, preventing the tires from getting any grip because it offers almost no traction, which makes accelerating, braking, and cornering dangerous tasks—like traveling on ice—especially for a single track road vehicle. Paradoxically, deep mud is preferred to filmy mud. This is because in deep mud, the mud surrounds the tires, thereby providing lateral support to keep them from sliding uncontrollably.

11.5 Noise Generation and Effect

The motorcycle rider is at the mercy of the elements, is directly exposed to the full force of the wind. The wind noise level increases with speed, so at high speeds the rider is subjected not only to increased drag but also to a high noise level. The main sources of the noise include the windshield, the helmet, and the body of the rider.

11.5.1 Windshield Noise Generation

The windshield is impacted directly by the head-on wind. This causes both vibration and deflection of the windshield, with the maximum deflection occurring at the top edge of the windshield (Fig. 11.7). The vertical position of the windshield substantially influences the level of noise that becomes attributable to the helmet and to the body of the rider.

In a high-post windshield, the wind deflected from the windshield traverses over the rider's helmet and over the rider's body, thus minimizing the noise contribution from both (Fig 11.7(a)). There is still noise contribution from recirculation and from side flow around the helmet and from the rider's body. There are, however, two disadvantages with the high-post windshield. The first is increased flutter. The higher the windshield, the more it approaches a vertical cantilever. It deflects more at the vertical edge, especially at high speeds. The greater the deflection, the greater is the flutter in the shield, and this transfers the longitudinal waves directly to the body of the rider. The second is the gap at the root of the windshield. A high windshield can be achieved either by raising the windshield high relative to the handlebar or by cutting a larger windshield material. If the former is done, the gap at the bottom of the windshield increases and air flows through this, directly hitting the front assembly and the rider's trunk. This results in increased drag, which may be substantial enough to undo the benefit of increased windshield height. A solution to this potential disadvantage is to maintain the root of the windshield below the handlebar, and sealed such that there is minimal gap for the head-on air to penetrate.

The low-post windshield directs the overflow to the rider. It is more sturdy than the high-post windshield, so it has less deflection and flutter. But due to its low top, relative to the rider, the overflow from the top edge combines with the direct air stream flow to hit the rider from the top of the helmet to as low as the rider's shoulder (Fig. 11.7(c)). The low-post windshield is not recommended for cruiser, touring, or standard motorcycles.

In the mid-post windshield, the airstream coming off the windshield just touches the top of the helmet. The deflection of the top of the windshield is less than for the high-post but more than that for the low-post configuration (Fig. 11.7(b)). It is considered ideal for a smooth road. A little bump can quickly change that benefit as the rider's face is momentarily in the airstream deflected from the windshield.

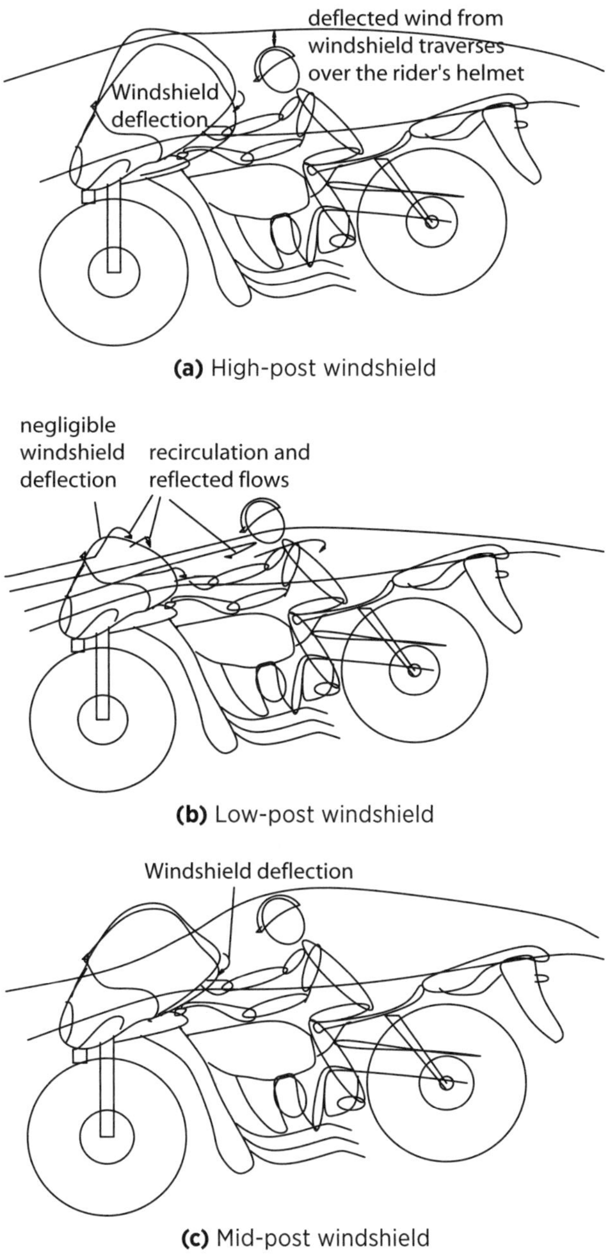

Fig. 11.7 Windshield deflection and flow perturbation

11.5.2 Helmet Noise Generation

The noise level from the helmet depends on the helmet material, the padding quality, the vertical position of the windshield top edge relative to the helmet, and the wind speed. As discussed, helmet noise will be greatest for a low-post windshield and least for a high-post windshield, all other factors remaining constant. For a given helmet material, the better the padding, the less is the noise experienced by the rider. It is important to note here that the helmet may generate the same noise intensity, but it is the reception by the rider's ear that is important. The better the padding, the less is the level of noise that is transmitted to the rider.

The wind speed and direction have individual and additive contributions to helmet-based noise. If the motorcycle is fitted with a high-post and wide windshield, the longitudinal wind speed has negligible effect on the helmet. But the cross wind, if present, affects the helmet. In this case, the higher the wind speed, and the closer the wind direction to being perpendicular to the temple, the greater is the helmet-based noise.

11.5.3 Body Noise Generation

Some noise is caused by direct wind or by air flow from the windshield hitting the frontal region of the helmet. But experiment has shown that most of the helmet-related noise comes from beneath the helmet, and that the root is air flow. Therefore attention should be paid to the flow around the neck of the rider because the flow velocity there is high [11-6]. The longitudinal flow below the shoulder is small compared with the flow in the neck and upper shoulder region. A wider windshield reduces even further the flow impinging on the mid-body, and fairings, when present, significantly reduce the flow on the lower body (Fig. 11.8).

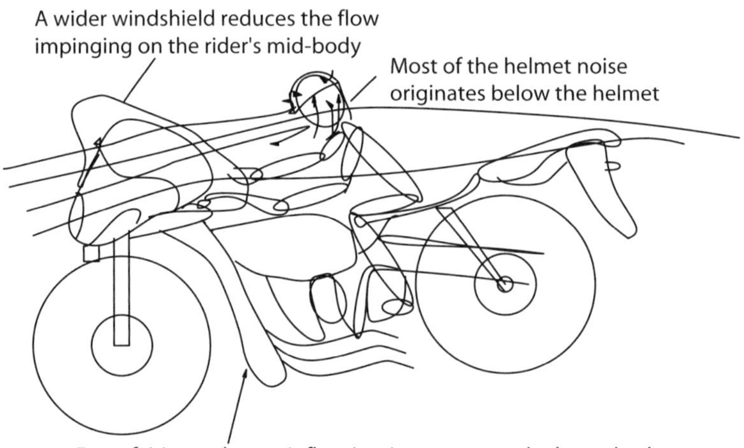

Fig. 11.8 Body noise generation and reduction

11.6 Aerodynamic Effects of the Track and the Pack

For a motorcycle, as for other road vehicles, a hard, well-paved terrain is more aerodynamically favorable than a soft terrain or one with loose top gravel.

The leading edge of the motorcycle and the windshield size and shape together affect the circulation behind the front rider. The following rider directly enters a suction region behind the leading vehicle. The benefit of this to a carefully positioned trailing vehicle is improved gas mileage. For this reason, a rider attempts to stay in second place behind the leading vehicle for most of a race, and then attempts to pass to the front toward the end of the race. The positioning is very important (Fig. 11.9). If the trailing vehicle is behind far enough for the circulation behind the leading vehicle to form a common front, the trailing vehicle will run into a high-speed bulk of air and will have to overcome the resulting high-pressure drag that may be worse than that experienced by the leading vehicle.

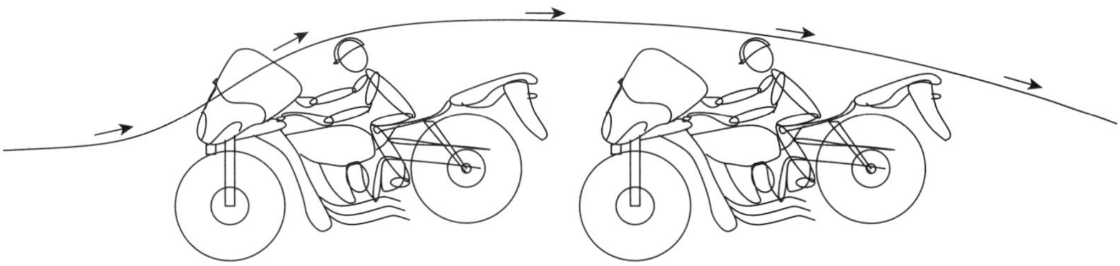

Fig. 11.9 Effects of the track and the pack

11.7 Motorcycle-Truck Interaction at High Speed

Motorcycle-truck dynamics are somewhat more complicated than car-truck dynamics because the lateral position of a motorcycle is not as precise as that of a car in the same lane. (A car lane can accommodate three motorcycles side-by-side.) When a motorcycle approaches the truck longitudinally, it is pushed away laterally from the truck, just as in the case of a car. The rider may feel the push so strongly that it appears as though the driver is being sheared away from the top of the seat. When the motorcycle is beside the truck, the rider can still feel a strong side force, due mostly to the side flow coming from the truck underbody and tire. The motorcycle swims in the turbulence from the truck's side flow throughout its travel beside the truck, including during its pull ahead past the truck. There is no noticeable lateral pull of the motorcycle toward the truck. Therefore, unlike the situation in car-truck interaction, the rider does not need a compensatory steer away from the truck when pulling away from the truck's environment (Fig. 11.10).

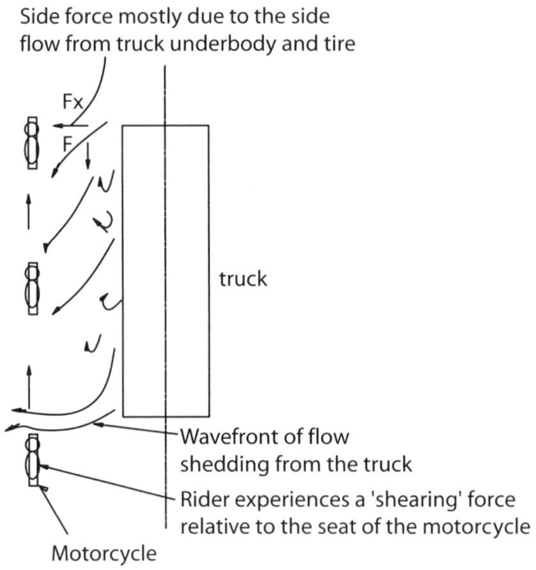

Fig. 11.10 Motorcycle-truck interaction

11.8 Motorcycle Engine Cooling

Many motorcycles are now liquid cooled because of better power cylinder performance and longevity. Traditionally though, motorcycles use air-cooled engines. Because such engines are exposed and highly finned, they are able to utilize the surrounding air for cooling. Forced cooling is effected by default as the speed of the motorcycle increases. The front fender, apart from reducing lift and enhancing downforce, directs the air flow such that the flow exits on the engine, thereby contributing to engine cooling (Fig. 11.11).

Fig. 11.11 Air-cooled engine

11.9 Three-Wheeled Motorcycle

Earlier three-wheeled motorcycles had the two wheels in the rear. They were generally called auto rickshaws (Fig. 11.12). Auto rickshaws were very valuable means of transportation of human and goods in many remote places in the world. Frontal two-wheel design is relatively modern (Fig. 11.13). A three-wheeled motorcycle has a higher stability than the two-wheeled counterpart in both the weave and the wobble modes [11-7].

Fig. 11.12 Auto rickshaw

Fig. 11.13 (MC-D300TKB) Modern three-wheeled motorcycle (reproduced with the permission of SAE International)

Summary

Motorcycles are not as fuel efficient as is generally perceived; the gross weight-to-fuel efficiency is lower than that of cars. Motorcycles are generally classified as cruisers, sport, touring, standard purpose, dual purpose, and dirt. Lift is not much of a concern (nor is the benefit of downforce realized) in a motorcycle because of its narrow design with very small planform.

Among the aerodynamic improvement devices on a motorcycle are the windshield, front fender, and fairings. Windshields are also functional for noise reduction and for safety.

Three-wheeled motorcycles with two frontal wheels are perhaps the latest in motorcycle varieties. They are more stable than the traditional two-wheeled motorcycles, but they do not have maneuverability of the two-wheeled motorcycles.

Key Terms

Auto rickshaw: Three-wheeled motorcycle with two rear wheels.

Engine fins: An engine fin is a heat transfer device on the external of an engine; fins are typically used in air-cooled engines.

Helmet noise: Noise generated from or transmitted by the helmet; helmet noise is a result of air interaction with the helmet, and it is affected by the air speed, the side gust, and the vertical position of the rider's head relative to the top edge of the windshield.

Review Questions

1. How would you validate a claim that a motorcycle is less efficient than a car?
2. A sport motorcycle is not recommended for long-distance travel. Why? Which types are recommended for long-distance travel?

Application Questions

Using sitter capacity as the criterion, which would you choose as a more efficient means of transportation: a passenger car or a touring motorcycle with a side seat? State your assumptions.

Exercises

1. Which is more fuel efficient, a 300 lb motorcycle that has a gas mileage of 35 mpg or a 450 lb motorcycle that has a gas mileage of 32 mpg? (Both motorcycles are two-sitters.)
2. Discuss and compare the stability of a rickshaw, a three-wheeler with two front wheels, and a standard two-wheeled motorcycle.

Project

Compare the overall drag for a two-wheeled motorcycle with those for a three-wheeled motorcycle with two front wheels, and a rickshaw. Make all three motorcycles of the same length and height, and without a windshield. What aerodynamic conclusion can you reach from the drag comparison?

References

1. Stermer, B., *Streetbikes: Everything You Need to Know*, Motorbooks Workshop/ MBI, 2006.
2. Garcia, N., Arocha, H., and Lorenzo, J, "Aerodynamic Study of Prototype Motorcycle Using Experimental Techniques and Flow Visualization," SAE Paper 2005-01-3993 E, 2005.
3. Raftery, B., "Determination of the Drag Factor of a Fairing Equipped Motorcycle," International Congress and Exposition, Detroit, February 27–March 2, 1995, SAE Paper 950197.
4. Piccioni, C. F., Cianetti, F., and Paolotti, A., "Numerical Investigations on the Front Fender of Motorcycle," International Congress and Exposition, Detroit, February 24–27, 1997, SAE Paper 970412.
5. Otombe, J., and Hasegawa, A., "Experimental Analysis of Sense of Stability in Motorcycle," Passenger Car Meeting and Exposition, Dearborn, Michigan, September 18–21, 1989, SAE Paper 891993.
6. Takahashi, Y., Kurakawa, Y., Sugita, H., Ishima, T., and Obokata, T., "CFD Analysis of Airflow Around the Rider of a Motorcycle for Rider Comfort Improvement," SAE Paper 2009-01-1155, 2009.
7. Sponziello, A., Frendo, F., and Guiggiani, M., " Stability Analysis of a Three-Wheeled Motorcycle," *SAE International Journal of Engines*, Vol. 1, Issue 1, Paper 2008-32-0062, 2008.

Chapter 12
Internal Aerodynamics and Cooling System

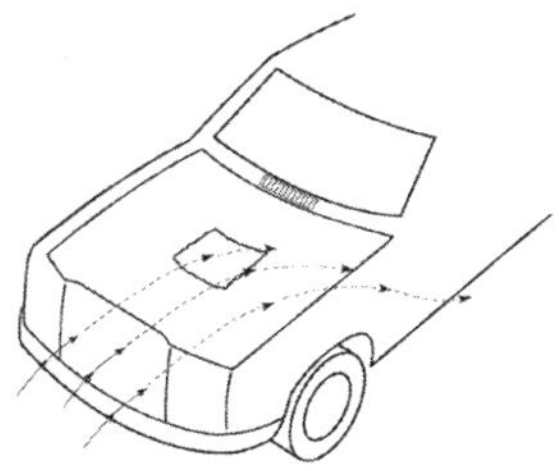

The study of the two main internal flow systems—engine cooling flow and passenger compartment or cabin flow—is essential to complete the study of ground vehicle aerodynamics. Even though we are primarily concerned with external flow, the internal flow behavior affects the external flow outcome. Because it takes away from the external flow, it reduces the drag, which is desirable. But because it exists, albeit within—rather than on the outside of—the vehicle, it introduces the undesirable internal aerodynamic drag. The internal flow, which is essentially concerned with the engine cooling system, the ventilation system, and (to a smaller extent) the brake cooling system, does not in general influence the external aerodynamic flow pattern, because it no longer sees the external flow until it exits at some point toward the rear of the vehicle. It may have some influence due to heat transfer from the underhood to the external flow to cause localized flow separation. In internal flow, mass of air is taken from the outside flow and directed to pass through a system of internal paths in the car. Historically, the primary design intent of the cooling system is (as the name implies) to cool the engine by effectively conducting the rejected heat from the engine and components to the environment. As more ways are sought to reduce drag and fuel consumption, greater attention has been given to drag reduction in the cooling system design. Although most of the aerodynamic drag on a vehicle is external, the contribution of internal flow to drag is not negligible. It has been estimated in some literature to be about 10–14% of the total aerodynamic drag [12-1, 12-2, 12-3, 12-4]. Even when it is considered minimal, it accounts for about 4% of the aerodynamic drag [12-5].

12.1 Underhood Initiated Underbody Flow

Unlike the external profile of a car, the underhood is a maze of components including block, hoses, and pumps, each serving a valuable purpose yet obstructing the effectiveness of other components (Fig. 12.1). The effective arrangement of this underhood potpourri is a measure of the skills and the coordination of the underhood design team.

It is important that cooling air flow is properly matched to the external flow, lest it contribute excessively to drag. Ideally, the exit speed of the engine cooling air should be greater than that of the underbody air flow. But the underbody air flow is highly dictated by the free stream velocity even after all the drag reduction features may have been incorporated into the underbody design. The practical approach is first to design the exit path of the engine cooling air such that the pressure of the exiting air is much lower than the pressure at the entrance

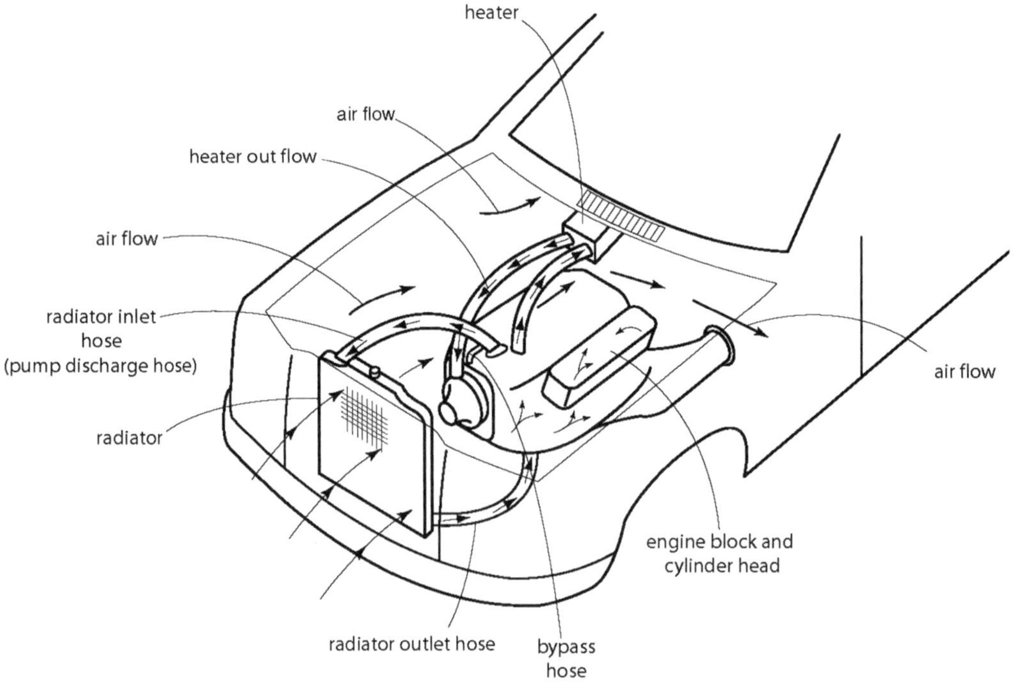

Fig. 12.1 Vehicle underhood

of the cooling path. This ensures that the flow is driven down the flow path to the exit in the underbody. Next, the speed of the exiting air should be higher than the design cruise speed of the car. This way, the exit speed from the engine cooling system is higher than the underbody air speed. The underbody air flow is then carried along by this plenum air. If the engine cooling air travels at a lower speed than the underbody flow, flow separation will occur both fore and aft of the point at which the engine cooling air enters the underbody flow, and a higher drag, as well as a possible increase in lift, will result (Fig. 12.2 7.4 (a),(b)).

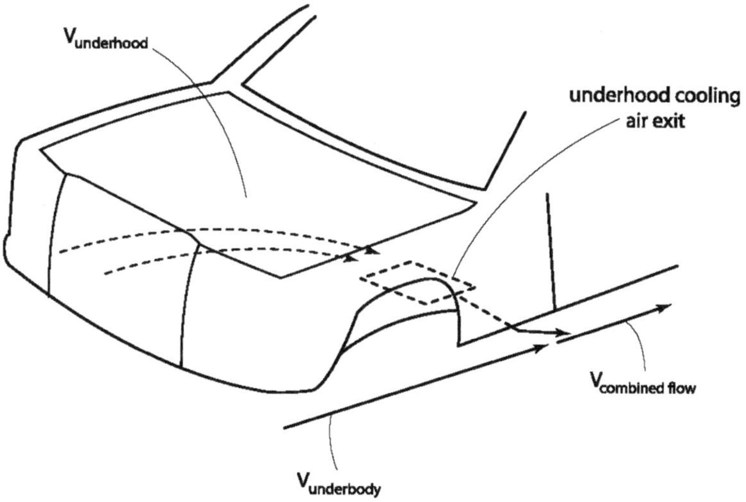

(a) $V_{underhood} < V_{underbody}$

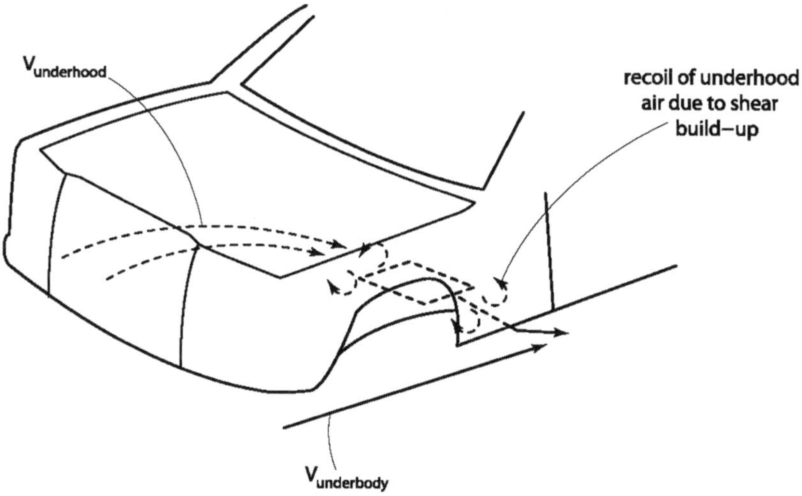

(b) $V_{underhood} > V_{underbody}$

Fig. 12.2 Interaction of exiting underhood air with underbody airstream

For inlet and outlet design, note the following guidelines:

- Place the inlet where the static pressure is greater than the ambient pressure.
- Place the outlet where the static pressure is less than the ambient pressure.

The pressure difference (Δp) drives the air flow in the duct or path of the flow.

12.2 Internal Air Flow

Although more common in race cars, modern passenger car designs have several ducts running within the vehicle—air ducts, ducts for electrical wiring, and brake line ducts (Fig. 12.3). Air is drawn in to provide secondary cooling (to radiators, oil coolers, and brake fluids) or primary cooling (to air-cooled engines and to the cabin). Establishing a flow profile for the internal flow in transport vehicles in general is not practical, but a simplified approach will suffice to provide an understanding of such flow and to reasonably quantify its contribution to aerodynamic drag.

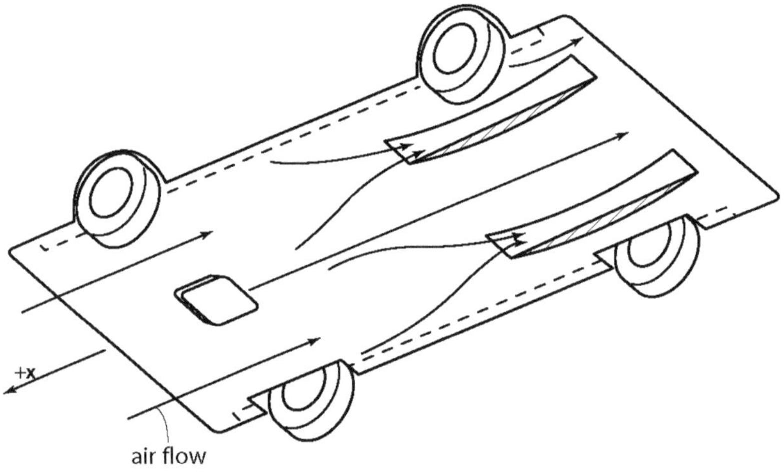

Fig. 12.3 Ducts in car underbody

12.2.1 Flow Through Pipe Systems

Unlike in the case of external flows, it would be incorrect to separate internal flows into the viscous layer and the inviscid region. Rather, viscosity influences the entire cross section of the flow. In Fig. 12.4(a), the shear layer starts to build at the wall and reaches a thickness d/2 at some point x_{max}. Unlike external flow, the viscous boundary layer is building from "both ends," and, more importantly, the pipe diameter d is in general too small to allow the flow to reach the inertia dominant region, so pipe flows are restrained to viscous condition. Figure 12.4(b) shows the extent, x'_{max}, that the viscous layer would have reached if the flow was on a free surface.

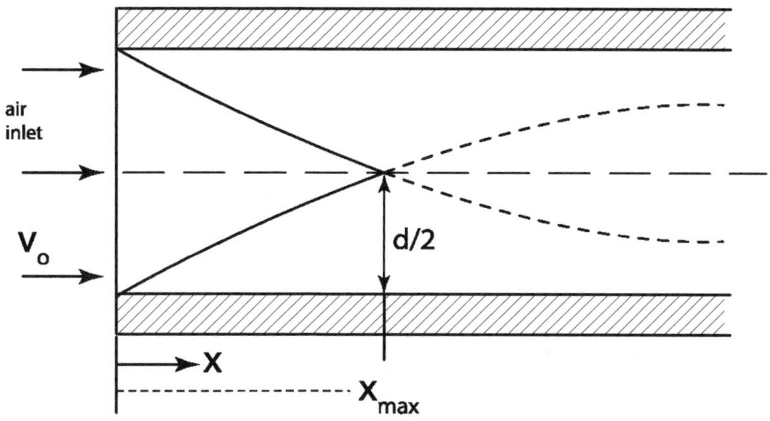

Fig. 12.4(a) Shear layer build-up in internal flow

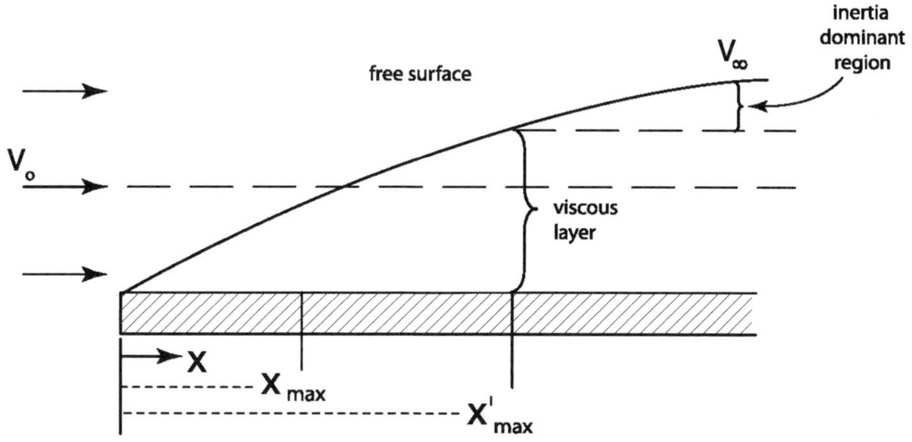

Fig. 12.4(b) Equivalent flow on a free surface

There are three main regions in a simple pipe flow: the entrance, the flow development region, and the fully developed flow region (Fig. 12.5(a), (b)). Typically, at the entrance region the flow is laminar and the velocity of the flow along the length of the pipe is practically the same as the relative free stream velocity (except at the wall of the pipe, where the velocity is zero). The flow is also uniform across the pipe diameter. In the second region, soon after the entrance, frictional effect of the wall causes a delay of the flow near the wall, resulting in the formation of a boundary layer that increases in thickness until it reaches the axis of the pipe in region 3. Since, from continuity, the flow rate is constant at any location in the pipe, the velocity of the fluid near the axis must increase to compensate for the wall effect. Therefore, the shape of the velocity profile changes accordingly until the flow is fully developed. The practical distance required to reach the final profile is referred to as the entrance length. Because the terminal velocity for turbulent flow is reached faster than for laminar flow, the entrance length is smaller than it is for a laminar flow through the same cross section.

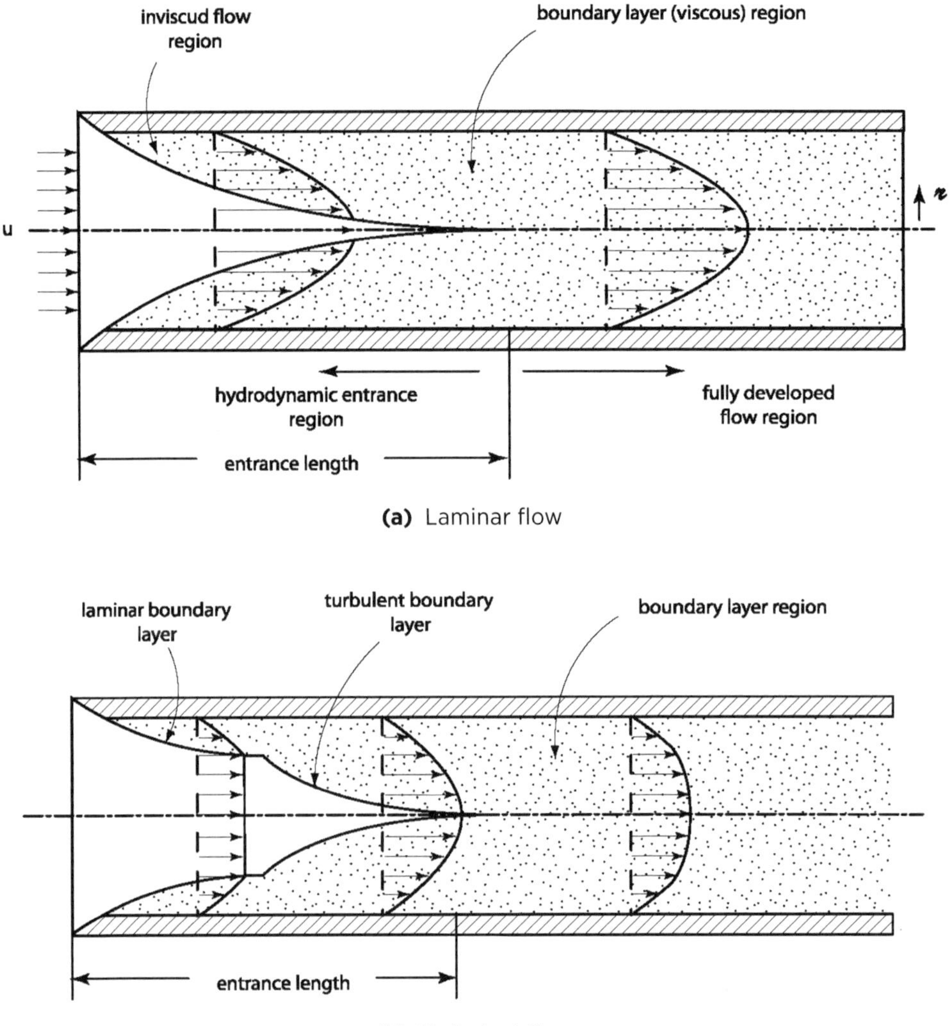

(b) Turbulent flow

Fig. 12.5 Development of boundary layer in a pipe

For the flow, the Reynolds number is given by

$$\mathrm{Re} = \rho \mathrm{vd}/\mu$$

From continuity

$$\rho_1 \mathrm{v}_1 \mathrm{A}_1 = \rho_2 \mathrm{v}_2 \mathrm{A}_2$$

The stagnation streamlined flow in a pipe is represented by Bernoulli equation

$$\mathrm{p}_0 = \mathrm{p} + \tfrac{1}{2}\rho\, \mathrm{v}_1^2$$

where

p_0 is the stagnation pressure at the entrance of the pipe;

p is the static pressure local to the pipe;

v_1 is the flow velocity local to the pipe; and

$\frac{1}{2}\rho v_1^2$ is the dynamic head.

Because of pipe friction, the pressure p_2 is lower than p_1 as the fluid travels in the pipe. Taking the pressure loss $\Delta p = p_1 - p_2$ into account, the Bernoulli equation is modified thus:

$$p_1 + \tfrac{1}{2}\rho v_l^2 = p_2 + \tfrac{1}{2}\rho v_2^2 + \Delta p \tag{12.1}$$

From continuity, $v_2 = v_1 = v$ for a straight pipe. So, as Eq. 12.1 shows, the summation of the static and the dynamic heads through a pipe is not a constant, even in a straight pipe, with the pressure loss having been caused by friction in the pipe. The smoothness of a pipe is generally rated in terms of its loss coefficient, ξ, defined as

$$\xi = \Delta p / (\tfrac{1}{2}\,\rho\, v_p^2) \tag{12.2}$$

where $v_p = u_{mean}$, the average fluid velocity in the pipe (usually taken as half the entrance velocity).

A force balance for steady flow in a horizontal pipe, Fig. 12.6 yields $(p_1 - p_2)\pi r^2 = \tau(2\pi r)\Delta L$, which, when simplified, yields

$$\tau = \Delta p r / 2\Delta L \tag{12.3}$$

where $\Delta p = p_1 - p_2$.

The basic Newtonian law of viscous resistance gives the shear stress as

$$\tau = \mu\, du/dr \tag{12.4}$$

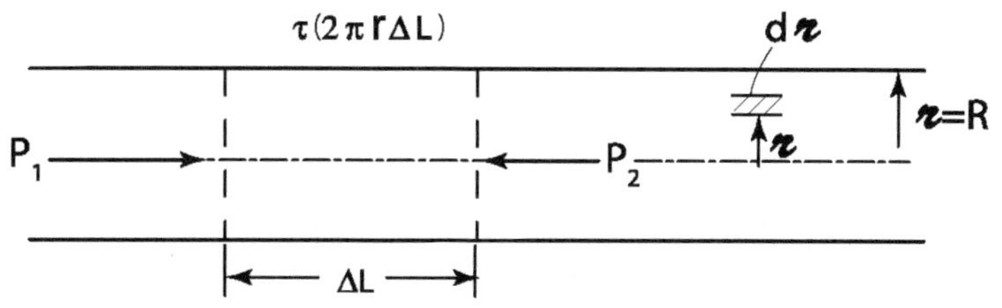

Fig. 12.6 Force balance for steady flow in a pipe

From (12.3) and (12.4)

$$\mu\ du/dr = \Delta p\ r/2\Delta L \tag{12.5}$$

Rearranging Eq. 12.5 and integrating with respect to r (since the flow profile varies only radially in a fully developed flow)

$$u = \frac{-1}{4\mu}\left(\frac{\Delta p}{\Delta L}\right)(R^2 - r^2)\Delta pr/2 \tag{12.6}$$

The maximum velocity, u_{max}, is obtained when r = 0 (i.e., at the axis of the pipe).

Hence, from Eq. 12.6

$$u_{max} = -\frac{R^2}{4\mu}\left(\frac{\Delta p}{\Delta L}\right) \tag{12.7}$$

The flow rate or discharge, Q, is defined by the product of the velocity and the area normal to the flow. Therefore, for an elemental strip, dr the discharge is

$$\delta Q = \frac{-1}{4\mu}\left(\frac{\Delta p}{\Delta L}\right)\left(R^2 - r^2\right)2\pi r dr \tag{12.8}$$

Integrating

$$Q = -\frac{\pi}{2\mu}\left(\frac{\Delta p}{\Delta L}\right)\int_0^R (R^2 r - r^3)dr \tag{12.9a}$$

$$= -\frac{\pi R^4}{8\mu}\left(\frac{\Delta p}{\Delta L}\right)$$

Equation 12.9a is generally referred to as the Hagen-Poiseuille equation.

If we replace Δp by $p_1 - p_2$, and ΔL by l, Eq. 9.9a can further be expressed as

$$Q = -\frac{\pi R^4}{8\mu l}(p_1 - p_2) \tag{12.9b}$$

Or, when written in terms of the pipe diameter

$$Q = -\frac{\pi d^4}{128\mu l}(p_1 - p_2) \tag{12.9c}$$

This expression is for laminar flow through circular, horizontal pipes, and this will suffice for even slightly slanting pipes, as are found in automobiles. The general expressions for inclined pipes (here stated for completion) are

$$u = -\frac{(R^2 - r^2)}{4\mu}\frac{\Delta(p + \rho gh)}{\Delta L} \tag{12.10}$$

and

$$Q = -\frac{\pi R^4}{8\mu}\frac{\Delta(p+\rho g h)}{\Delta L} \tag{12.11}$$

where h denotes the location of the fluid element from a reference height.

The mean velocity is simply Q/(cross sectional flow area).

Recall Eq. 12.9a, $Q = -\frac{\pi R^4}{8\mu}\left(\frac{\Delta p}{\Delta L}\right)$, and the area of flow πR^2. The mean velocity is obtained from

$$u_{mean} = -\frac{\pi R^4}{8\mu}\left(\frac{\Delta p}{\Delta L}\right)/(\pi R^2) = -\frac{R^2}{8\mu}\left(\frac{\Delta p}{\Delta L}\right) = -\frac{D^2}{32\mu}\left(\frac{\Delta p}{\Delta L}\right) \tag{12.12}$$

From (12.7) and (12.12) we see that the mean velocity is half the maximum velocity:

$$u_{mean} = \tfrac{1}{2}\, u_{max} \tag{12.13}$$

In general, when given the pipe entrance velocity, the average velocity is merely expressed as half the entrance velocity (as indicated above in the expression for vp).

For Reynolds numbers below 2300, the flow is generally laminar, and the head loss in a relatively smooth pipe is given by the Darcy-Weisbach expression

$$h_\xi = \xi\,\frac{\Delta L}{D}\frac{u_{mean}^2}{2g} \tag{12.14}$$

with $\Delta p = \gamma h_\xi$, u_{mean}, in Eq. 12.13 can be rewritten as $u_{mean} = \frac{D^2}{32\mu}\left(\frac{\gamma h_\xi}{\Delta L}\right)$ which, when solved for h_ξ yields

$$h_\xi = \frac{32\mu}{\gamma}\frac{u_{mean}\Delta L}{D^2} \tag{12.15a}$$

When rearranged, the last expression becomes

$$h_\xi = \frac{64}{[\rho u_{mean} D/\mu]}\frac{u_{mean}^2}{2g}\frac{\Delta L}{D} \tag{12.15b}$$

Setting Eq. 12.14 equal to Eq. 12.15b, and noting that the expression in bracket denotes the local Reynolds number, Re_l, we have

$$\xi\,\frac{\Delta L}{D}\frac{u_{mean}^2}{2g} = \frac{64}{Re_l}\frac{u_{mean}^2}{2g}\frac{\Delta L}{D}$$

which gives

$$\xi = \frac{64}{\mathrm{Re}_l}$$

This equation, which plots as a straight line with slope –1 on a log-log chart (Fig. 12.7), is generally used for all laminar flows in pipes because the head loss in laminar flow is independent of the roughness of the pipe wall.

Beyond Reynolds numbers of 2300, the flow is turbulent, and the pipe wall roughness becomes significant in its contribution to the head loss. Blasius was the first on record to correlate the smooth-pipe experiments in turbulent flow [12-6]. He presented the results by an empirical formula that is valid up to a Reynolds number of 10^5. The Blasius formula is

$$\xi = \frac{0.3164}{\mathrm{Re}^{1/4}}, \qquad 2300 < \mathrm{Re} < 10^5$$

Beyond a Reynolds number of 10^5, the relationship between friction factor and Reynolds number is

$$1/(\xi^{1/2}) = 2\log(\mathrm{Re}\,\xi^{1/2}) - 0.8, \qquad \mathrm{Re} > 10^5$$

The rougher the pipe, the greater the drag and the less the variation of the pipe friction coefficient with Reynolds number. The term $\frac{\varepsilon}{D}$ in Fig. 12.7 is called the relative roughness. In most of the internal flow problems for pipe and channel flows in ground vehicles, the pipe roughness will hardly be of significance because the flow paths are in general relatively smooth.

12.2.2 Curved Pipes

Curved pipes and ducts are inevitable components of ground vehicles. They are therefore worth our aerodynamic consideration, even if briefly. Curved pipes are liable to have flow separations (Fig. 12.8). Losses occur when the flow is caused to change its direction. When fluid flows in a curved path, a radial force acts on the fluid to provide the centripetal acceleration. As a result, the pressure is increased near the outer wall of the bend, and there is a reduction of pressure near the inner wall. This results in an adverse pressure gradient, with the pressure increasing in the direction of flow, and a subsequent flow separation. In the case of curved pipes, there is additional drag due to the pipe friction. This drag, along with such drag contributors as elbows, joints, and valves in pipelines, is usually referred to as minor head loss, although (and ironically so) the loss is sometimes greater than that due to the pipe friction. The probability of flow separation in a curved pipe increases with the pipe bend, γ, and decreases with the curvature radius, r, and the pipe inside diameter, d (Fig. 12.8).

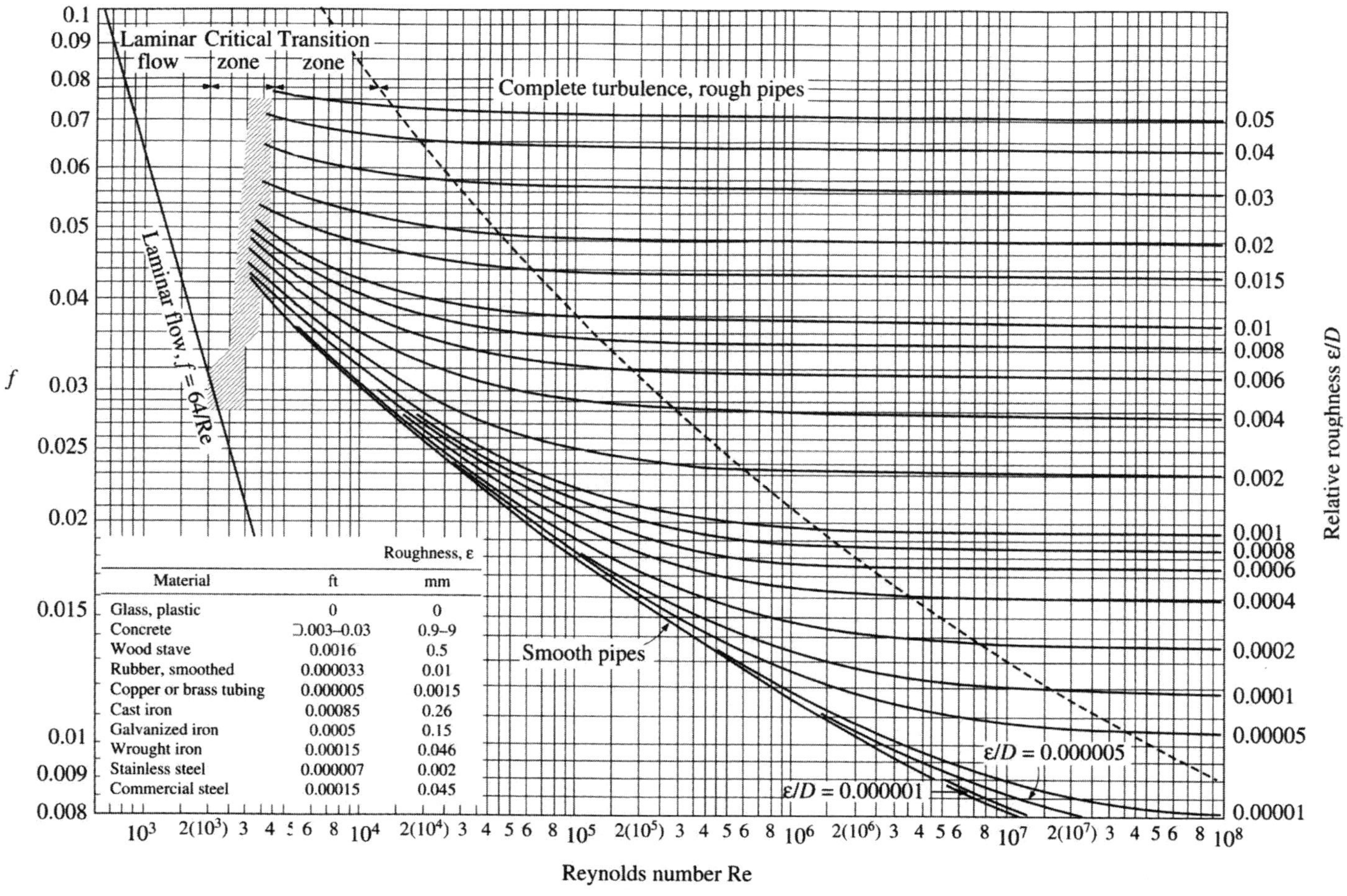

Material	Roughness, ε ft	Roughness, ε mm
Glass, plastic	0	0
Concrete	0.003–0.03	0.9–9
Wood stave	0.0016	0.5
Rubber, smoothed	0.000033	0.01
Copper or brass tubing	0.000005	0.0015
Cast iron	0.00085	0.26
Galvanized iron	0.0005	0.15
Wrought iron	0.00015	0.046
Stainless steel	0.000007	0.002
Commercial steel	0.00015	0.045

Fig. 12.7 Moody diagram

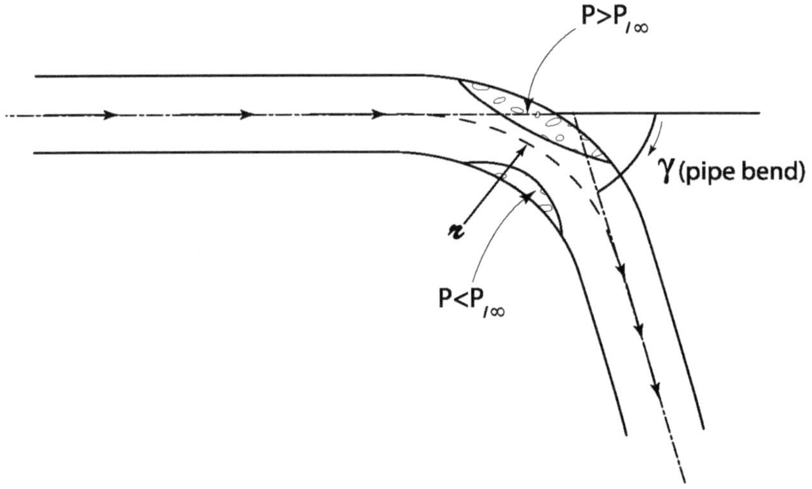

Fig. 12.8 Flow in a pipe bend

We have just idealized the internal flow in ground vehicles. In reality, the flows are more complicated. The inlets are usually blunt rather than gradual, and abrupt flow restrictions and enlargements are common features (Fig. 12.1). When the inlet is sharp, the loss coefficient can be as much as 10 times the loss from a rounded inlet. An abrupt flow restriction results in flow separation before and after the restriction, and a sudden enlargement causes flow separation at the immediate entrance of the larger section, with the result that the pressure rise is reduced (Fig. 12.9 (a), (b)).

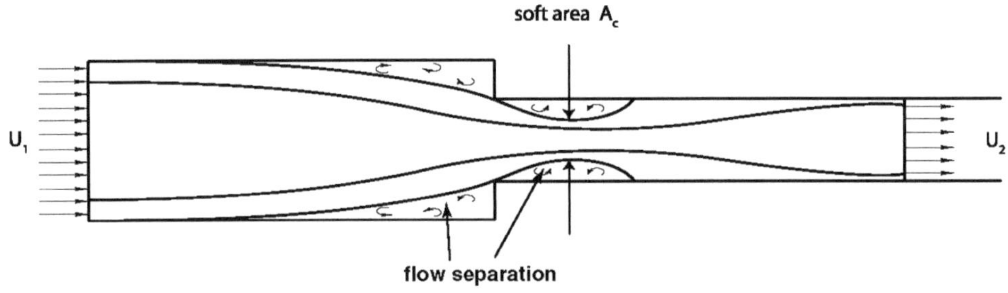

(a) Abrupt contraction

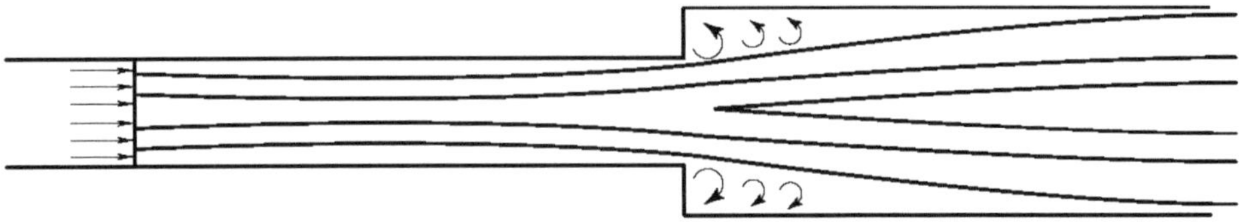

(b) Abrupt expansion

Fig. 12.9 Changing pipe cross section

12.3 Radiator Air Flow

The most significant internal flow drag contributor is the flow through the radiator. Apart from the cooling drag penalty resulting from cooling air flow, the vehicle nose lift and the side forces due to cross wind also increase as the cooling drag increases [12-7]. As air is channeled through the radiator (as well as other components, such as the oil cooler), it is desirable to maximize the air flow rate at the least possible drag. A radiator slanted forward appears to be the best orientation to achieve such a goal [12-8]. It is not practical, however, because the hot air released at the outlet would flow straight into the air pool for the vehicle's passenger compartment fresh air intake. The arrangement nonetheless serves as a guide in the positioning of the radiator. A more practical design directs the flow to exit mostly just below the front seat location, where it infuses into the underbody flow, and partly further downstream along the longitudinal ducts. Some may escape, particularly at lower speeds, at the hood-windshield interface, depending on the level of sealing (Fig. 12.10).

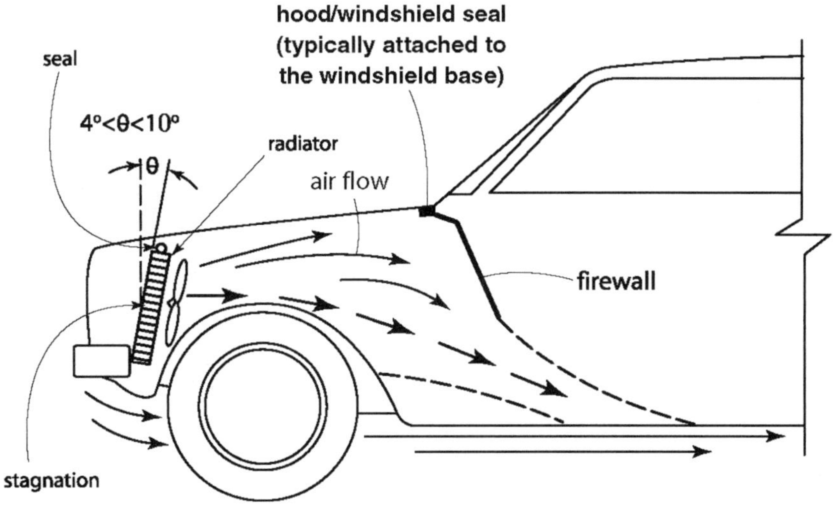

Fig. 12.10 Optimum radiator position

The fan and shroud arrangement is important not only to the cooling effectiveness of the radiator assembly but also to the internal aerodynamic drag contribution. The shroud can be the venturi type, ring type, or box type (Fig. 12.11), and the fan design can be axial, centrifugal, or mixed flow (Fig. 12.12). In the venturi type shroud, the fan is completely immersed in the shroud, with the latter having a tangency point at around two-thirds of the fan's projected width, and from there it diverges to direct the radial component of the air around the engine components. Both the ring type and box type fan shrouds direct the air flow head-on to the engine, with the ring type reaching a broader first-contact area of the engine.

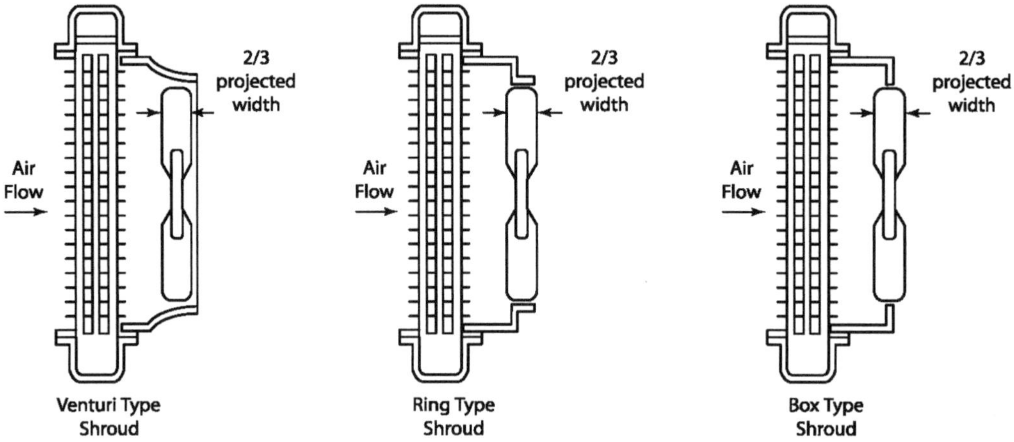

Fig. 12.11 Radiator-shroud arrangement

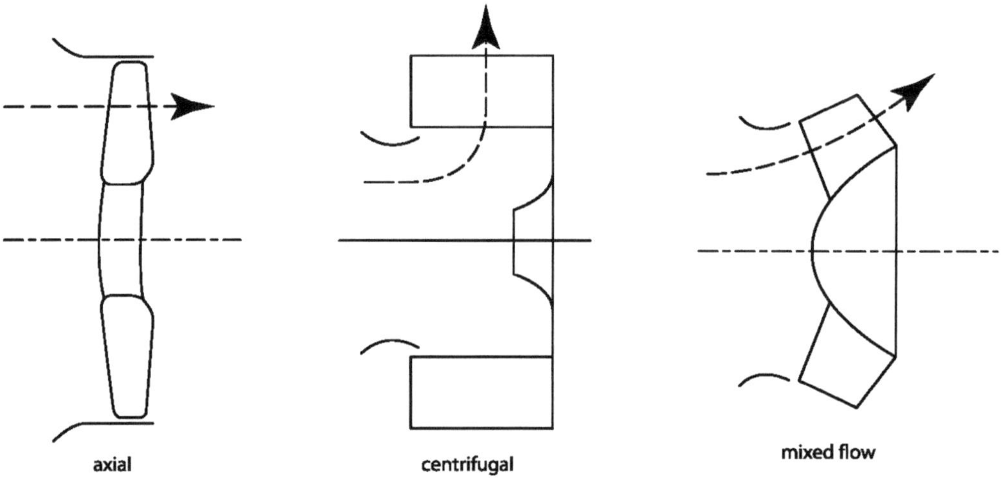

Fig. 12.12 Types of radiator fan

12.3.1 Cooling System Drag

There is no precise way to determine the cooling system drag, but an approximate approach is to measure the overall drag of the vehicle in normal operation. The radiator assembly is then blocked off, and the obtained drag is subtracted from the earlier measured value. Actually, the resulting drag assigned to the cooling system is an accumulation of the drag due to the cooling system and the interaction between the cooling system air flow and the main flow. The higher the vehicle speed, the greater is the significance of this interference drag. A more accurate approach is to measure the drag contribution from the cooling system directly with a strain-gage force balance system or some other electronic device [12-9, 12-10]. With a pressure probe at different locations in the front of the vehicle and under the hood, the relative values of the pressure coefficient can be obtained, and the stagnation and low air flow areas can be determined. Then the necessary incremental changes to the shape of the front-end components—bumper, hood, fascia—are made ultimately to reduce the aerodynamic drag. The same approach applies to the radiator cooling air, which travels from the engine block, engine components, and attachments. Air dams help to infuse air into the underhood and the engine floor, thus increasing the air volume. Axial flow fans are also most effective in increasing the air flow volume without the accompanying pressure increase experienced when the centrifugal flow fan is used.

Because of the limited frontal area historically made available for cars and off-highway vehicles, the surface area of the radiator is much greater than the frontal area of the vehicle. The air flow across a vehicle cooling system naturally experiences a drop in pressure because of the system resistance, and a considerable surface friction drag results from the flow through the radiator, even at low speed. Grille requirement is linked to the shape of the front end and to the underhood air flow volume. Although grilles enhance the look of a car, they generally add to the drag of the internal air flow (Fig. 12.13). It is therefore essential to evaluate the effectiveness of the fascia concurrently with the front end and the underhoood air flow analysis. Mild alteration in the front end affects the inlet cooling air path, and the design and positioning of engine and components under the hood will also affect the air path and the air flow through the radiator. The less the restriction to air flow, the less is the incurred drag, and the better the aerodynamic performance of the vehicle.

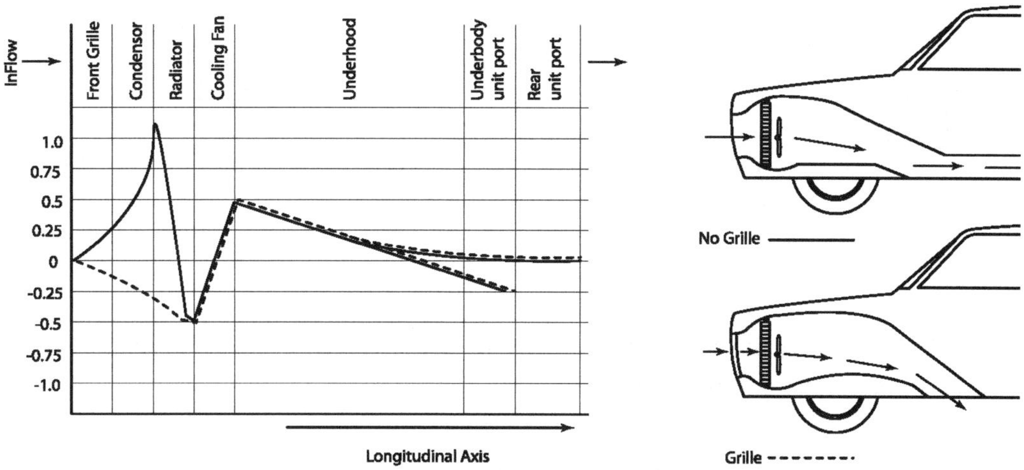

Fig. 12.13 C_p versus component location

The size and the style of the exit port are probably the most crucial in the cooling system air flow design. We will here develop a benchmark for sizing the exit port (Fig. 12.14).

From (12.2), we can express the pressure drop across the radiator as

$$\Delta p = (\tfrac{1}{2}\,\rho\, v_c^2)\xi$$

where v_c is the average air flow into the core, and ξ is the pressure loss (or friction) coefficient. At low speeds and laminar flow, ξ is proportional to $1/\, v_{c'}$, and at high speed when the flow is turbulent, it is proportional to $1/v_c^{1/5}$. The more cooling fins for a given radiator architecture, the greater is the value of ξ.

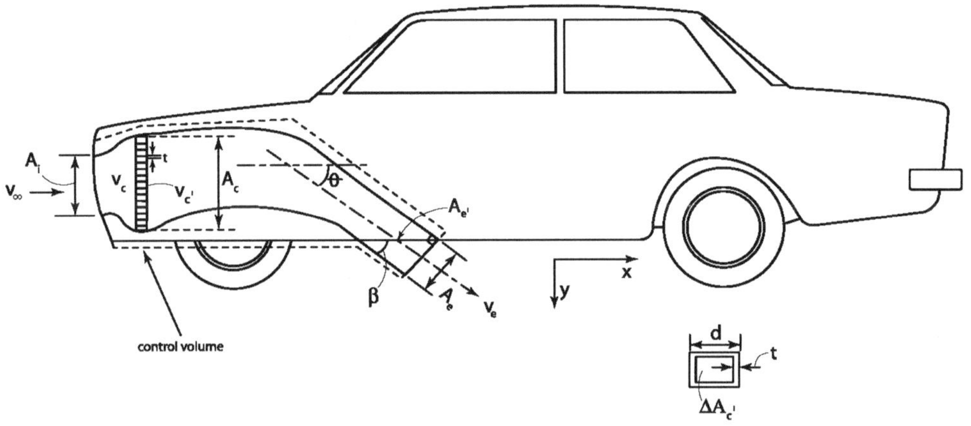

Fig. 12.14 Internal flow

It is important to understand that v_c is the velocity just before the core rather than the velocity, $v_{c'}$ in the core. How significant is the difference? Let the thickness of the core fins be t, surrounding an air flow cross section $\Delta A_{c'}$. If we assume a square cross section of side d, then$\Delta A_{c'}$ is $(d - 2t)^2$. For n such cross sections in the radiator, the total air space available is $n\Delta A_{c'}$, and the solid area,$(Ac - n\Delta A_{c'})$ is $n\{2dt + 2(d - 2t)t\}$ or $4nt(d - t)$. Since n is usually in the hundreds, this can be relatively significant.

From continuity, $v_c A_c = v_{c'}(n\Delta A_{c'}) = v_{c'}\,\{A_c - 4nt(d - t)\}$.

The last expression clearly indicates that $v_{c'}$ is greater than v_c. Continuity is therefore maintained by using v_c since we are employing the radiator core area, A_c. Besides, it is much more practical to measure v_c than to determine an average $v_{c'}$ or a corrected area, $A_{c'}$ (i.e., $n\Delta A_{c'}$).

The outlet area is $A_{e'}$ but for ease of analysis we shall use A_e, the face normal to the exit flow, and correct for $A_{e'}$ using the relationship $A_e = A_{e'}\sin\beta$ as shown in Fig. 12.14.

$$\text{Continuity: } v_c\, A_c = v_e\, A_e$$

Rewriting (12.1), and substituting for Δp,

$$p_\infty + \tfrac{1}{2}\,\rho\, v_\infty^2 = p_e + \tfrac{1}{2}\,\rho\, v_e^2 + \tfrac{1}{2}\,\rho\, v_c^2 \xi \tag{12.16a}$$

Replacing v_e with $v_c\, A_c\, /A_e$ from continuity, (12.16a) becomes

$$p_\infty + \tfrac{1}{2}\,\rho\, v_\infty^2 = p_e + \tfrac{1}{2}\,\rho\, v_c^2 (A_c/A_e)^2 + \tfrac{1}{2}\,\rho\, v_c^2 \xi \tag{12.16b}$$

Rearranging Eq. 12.16b yields

$$\frac{v_c}{v_\infty} = \sqrt{\frac{1 - C_{pe}}{(A_c/A_e)^2 + \xi}} \tag{12.17}$$

where $c_{pe} = \dfrac{p_e - p_\infty}{\frac{1}{2}\rho v_\infty^2}$, the pressure coefficient at the exit region.

Now, taking the summing the cooling forces in the control volume and equating the *x*- and *y*-components to the respective linear momentum, assuming steady flow, we get

$$D_c = p_\infty A_i - p_e A_e \cos\theta + \dot{m}(v_\infty - v_e \cos\theta) \tag{12.18}$$

and

$$L_c = p_e A_e \sin\theta + \dot{m} v_e \sin\theta \tag{12.19}$$

where

L_c is the contribution to lift;

D_c is the drag force;

□ is the air mass flow rate, $\rho\, v_c\, A_c$, for the steady, incompressible flow; and

A_i is the inlet cross section of flow (i.e., the grille open area).

Equations (12.18) and (12.19) are normalized when divided by $\frac{1}{2}\rho\, v_\infty^2\, A_f$, where A_f is the vehicle projected frontal, to give the drag coefficient and the lift coefficient due to flow of cooling air.

$$C_{Lc} = C_{po}\frac{A_e}{A_f}\sin\theta + 2\left(\frac{v_c}{v_\infty}\right)^2 \frac{A_c}{A_f}\,\frac{A_c}{A_e}\sin\theta \tag{12.20}$$

and

$$C_{Dc} = \frac{p_\infty A_i - p_e A_e \cos\theta}{\frac{1}{2}\rho\, v_\infty^2 A_f} + 2\frac{v_c A_c}{v_\infty A_f}\left(1 - \frac{v_c A_c}{v_\infty A_e}\cos\theta\right) \tag{12.21a}$$

The expression on the left of Eq. 12.21a can be further refined using continuity relationship $p_\infty A_i = p_e A_e$, and $\frac{v_e}{v_\infty} = \cos\theta$ (Fig. 9.14) to give

$$-\frac{(p_e - p_\infty)}{\frac{1}{2}\rho\, v_\infty^2}\frac{A_e}{A_f}\cos\theta$$

or

$$-\,C_{pe}\frac{A_e}{A_f}\cos\theta$$

We can then write (9.21a) as

$$C_{Dc} = -C_{pe}\frac{A_e}{A_f}\cos\theta + 2\frac{v_c A_c}{v_\infty A_f}\left(1 - \frac{v_c A_c}{v_\infty A_e}\cos\theta\right) \qquad (12.21b)$$

The left term in Eq. 12.21b is the contribution due to the pressure at the exit port. It is already included in the external drag measurement by default and so should not be added to the cooling drag contribution when the latter is itemized in addition to the measured external drag. The second term is the drag due to the momentum in the vehicle longitudinal direction. The second expression of the second term is the drag reduction due to the slant of the exit port in the axial direction. The closer the inclination of the outlet air stream to the vehicle floor, the less is the drag contribution. It shows that the maximum reduction in drag and, from Eq. 12.20 the least contribution to lift is a design with an exit port straight to the back (i.e., $\theta = 0$). This is why race cars tend to have the exit port at the rear of the car (Fig. 12.13). Similarly, it shows that a vertical exit port design would give at least beneficial cooling drag as well as maximum lift contribution, which is unfavorable to vehicle stability. Also, since air exits rather vertically downward from the engine area, the flow is normal to the vehicle underside and results in a high cooling drag. However, it contributes to the downforce of the vehicle when the air exits upward from the hood. But, as seen in Eq. 12.21 it also adds to the drag. So, a balancing act—the art of engineering—is required between reducing the drag and enjoying more downforce.

In general, in spite of the downforce benefit that such a design may produce, it is not recommended. Apart from the inherent cooling drag contribution, such a design would cause additional drag due to flow separation just before the windscreen, as well as the impracticality of ejecting hot exhaust air more or less in the driver's face.

There is also the temptation to have the cooling air outflow on the sides of the vehicle, as is sometimes done. Though this has no effect on the lift, it contributes to the skin friction drag as well as the interference drag. It may also (albeit slightly) result in an increase in the vehicle frontal area. Although the desirable longitudinal flow exit suffers momentum loss due to friction, the loss is reduced by ensuring a high exit velocity. The combination of a small exit area and a large inlet (radiator) area ensures this outcome. A_i, the grille opening area, becomes relatively unimportant beyond the extent of sizing it, so that there is no flow separation along the diffuser as the flow travels to the radiator.

12.3.2 Sizing Cooling System Port

Recall Eq. 12.17, and write the left term as $\dfrac{v_c/v_e}{v_\infty/v_e}$.

v_∞/v_e is practically a constant, K. We can therefore write Eq. 9.17 as

$$\frac{v_c}{v_e} = K\sqrt{\frac{1-C_{pe}}{(A_c/A_e)^2+\xi}} \tag{12.22}$$

From Eq. 12.22 we can see that a large A_c/A_e translates into a high exit velocity, and a change in the area ratio produces a second order change in the velocity. Equation 12.22 is the benchmark for sizing the cooling system ports.

Most cooling systems are able to provide the needed air flow with just the ram air. The radiator fan is usually required during idle or at low speed—a common phenomenon in city traffic. The electric fan has more or less displaced the one-time engine-coupled cooling fan (Fig. 12.15). Because the electric fan is independent of the engine shaft rotation, it can be switched on only when needed. This transition saves the power that would have been wasted from unnecessary running of the fan at highway speeds.

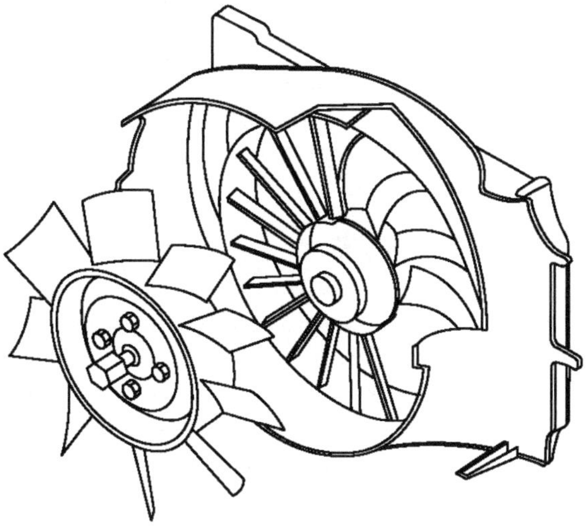

Fig. 12.15 Electric fan (with shroud)

12.4 Thermal Effect

Air entrained in the internal flow path rises substantially in temperature from the time it reaches the radiator to the time it exits aft of the engine. This higher temperature outflow affects the external flow both directly and indirectly. The direct effect on the external flow is experienced when the air entrains in the underbody air stream just aft of the engine compartment. The temperature of the combined flow is suddenly higher than the temperature of the flow entering the facia grille, the same that is maintained in the underbody up until the exit point of the air from the cooling system. The air rises, thereby adding to the lift on the vehicle. Indirectly, the heated air in the ducts and around the engine causes the temperature of the contact face—the hood, the duct walls—to rise. This in turn conducts the higher temperature to the outer face. Thus we have a contribution to the hotness of such surfaces as the hood and the cabin floor. A higher hood exterior temperature favors flow separation and added induced drag. A higher cabin floor temperature requires greater supply of cooling air in the cabin for passenger comfort (although this would be a benefit in the winter when the cabin needs heat for passenger comfort). It is therefore necessary to duct as much air as feasible to reduce the air exit temperature, most of which is generated aft of the radiator.

Although the momentum of the internal air flow in the direction of motion is generally low, this relatively hot air contributes to the lateral flow of air in the underside of cars, thus adding to the volume and the turbulence of the upstream flow experienced by the front wheels. The term interference drag is used to describe the spreading effect of the cooling air on the mainstream flow in the underside of the car. This effect adds to the tire drag. The cooling-air drag, as it is called, is strictly the drag that results from the pressure loss in the cooling air duct. It is generally used, however, to include the interference drag; this is understandable because it is complicated to attempt to separate the interference drag from the strictly cooling-air drag. Slight successes at such an attempt have shown the insignificance of the interference portion compared with the duct portion, and hence the impracticality of the attempt to differentiate between both.

12.5 Cabin Air Conditioning

The effect of cabin air flow on aerodynamic drag is probably the least significant, yet it is not negligible. The significance of the ventilation system (also referred to as passenger air comfort system or cabin air distribution system) is in the selection of the air inlet and outlet ports, primarily to utilize the existing pressure distribution and pressure gradient to enable air flow. Cabin air flow is defined as the flow in the cabin when all windows (and, where applicable, the sunroof) are closed. Internal comfort affects the driver's performance and alertness, both of which affect safety on the highway. Constant air exchange is necessary as the vehicle drives along to ensure the freshness of air intake by the occupants. The recommended rate of air intake in a passenger car is 1060 ft^3/h (30 m^3/h). The smaller the volume of the cabin, the greater the required number of air changes per unit time, and the greater the power demand from the cabin air control unit, be it the heater or the air conditioner.

For passenger health and comfort, air in the cabin needs to be clean and dry. Clean air is free of particles and all forms of pollution—physical, chemical, and noise pollutions. The requirement of clean air naturally dictates that the air inlet be above the height of the combustion air. Hence the vertical location needs to be above the hood. Because it cannot interfere with visibility, it needs to be located below the windshield. With such upper and lower constraints, the vertical location of the cabin air inlet is generally in the region above the hood and below the windshield.

The longitudinal and latitudinal coordinates of the location can be tricky. First, one seeks a point of maximum (or, at least, very high) static pressure so that the air can be self-driven through the cabin to the air outlet (see Section 12.1). Then, based on soiling test—because hood soiling affects the pressure at the inlet air—the middle or the front corners are selected at the height already imposed in the vertical position determination. A popular location for the fresh air intake into the cabin is in the front, around the axial location of the engine firewall. The outlet is usually creviced in the dashboard assembly. In response to human physiology, hot air is supplied through outlets that have been designed to graduate the air temperature along the vehicle vertical axis such that the temperature experienced at the feet is the highest and the lowest is at the face. Air flow around the vehicle occupant can help to provide a cooling effect in a hot environment. This is often referred to as factory air in cars that are not equipped with an air conditioning unit. The cooling effect is the same that a person experiences from a fan, rather than an air conditioner, on a hot summer day.

The location of the air exiting the vehicle air outlet is arguably not as important as that of the air inlet for a number of reasons including passengers constantly open the doors, passengers winding down the windows, and natural leaks in the vehicle. Nonetheless it still calls for some attention. The location of the air outlet from the vehicle should be a relatively low pressure area and should be aft of the passenger compartment. Such locations include just behind the D-pillars, below the D-pillars, in the rear window just before the low end of the windshield, and within the rear bumper. Designers use any of these or a combination thereof. There has been no consensus on the best location.

Although low humidity is undesirable, it can be tolerated. Not so with high humidity (>7%); an air conditioner becomes essential in that instance. A price comes with the comfort that an air conditioning unit provides. For example, part of the output power (>2 kW) is diverted to running the a/c compressor. This translates to a higher fuel demand than there would have been without the air conditioner.

The cabin air distribution system includes the inlet cowl, inlet plenum, a/c registers, and defrost ducts. Limited space is available under the instrument panel, which is where the cabin air distribution system is usually positioned. An ideal duct system is quiet in operation and provides maximum specific volume of air flow at minimum drag penalty. Established computational methods are employed as soft design tools to optimize the flow in the basic duct design for a vehicle.

The inlet plenum has openings that connect to the blower housing inlet and to the ventilation inlets. Care must be taken in its design to ensure that there are no openings for underhood air or gases to get into the HVAC system. The blower pumps the treated air through the ducts into the vehicle cabin for occupant comfort. The larger the blower, the greater its efficiency, and the quieter is the operation. The goal is to use as large a blower as the footprint will allow. A blower operating at about 3000 rpm is generally adequate. The blower assembly should generally be placed upstream of the heater core. With this arrangement, the noise experienced by the passenger is minimal, and the blower is better able to cool the warm air produced as air passes over the blower motor.

Ducts—adapter ducts, defrost ducts, heater floor ducts—serve to transfer air to the desired location within the vehicle. Adapter ducts, for example, connect various sections (such as the heater duct, panel duct, and defrost) of the system, just as an elbow or a joint in a pipe. The cross section area of a duct must be no less than the outlet to which it delivers air to avoid acceleration of flow, which results in rush noise and air flow resistance (Fig. 12.9(a)). Noise translates to passenger discomfort, and flow resistance contributes to internal flow drag [12-11].

Summary

Internal aerodynamics for ground vehicles consists, in essence, of engine cooling flow and cabin air flow. It increases vehicle internal aerodynamic drag, which can be as much as 14% of the total aerodynamic drag.

The air duct serves primarily to draw cooling air to the engine and the passenger compartment. Additionally, it provides secondary cooling for the radiator, oil coolers, and brake fluids. The largest duct through the vehicle is that for engine cooling. Design the exit path of the engine cooling air to have an exit air pressure that is much lower than the inlet pressure so that the flow can be naturally driven down the flow path to the exit in the underbody.

The ram air typically suffices to provide the needed engine cooling, especially in highway driving. In city traffic—low speed and idle conditions—ram air is minimal. Therefore, an assist in the form of a fan is required for continuous engine cooling. The electric fan has almost entirely replaced the engine-coupled cooling fan. Because the electric fan is independent of the engine shaft rotation, it can be switched on only when needed. This transition saves the power that would have been wasted from unnecessary running of the fan at highway speeds.

It is essential to duct as much air as feasible to reduce the air exit temperature, most of which is generated aft of the radiator. Unducted or poorly ducted air can result in an increase in the surface temperature of the hood and of the cabin floor. A higher hood exterior temperature favors flow separation and added induced drag. A higher cabin floor temperature requires a greater supply of cooling air in the cabin for passenger comfort.

Key Terms

Air duct: A channel (open or closed) that directs the flow of air.

Cabin floor: The floor space below vehicle occupants.

Engine compartment: The space, usually under the hood of a vehicle, where the engine is paced.

Internal flow: The flow within the vehicle compartment; it is not exposed to the vehicle profile.

Passenger compartment: Also called the cabin (in trucks); this is the space occupied by passengers in a vehicle.

Underhood: The space under the hood of a vehicle; it is the most common location of the engine, the radiator, the water pump, and other engine auxiliaries.

Review Questions

1. Mention a benefit and a disadvantage of internal air flow in ground vehicles.
2. Which two vehicle systems are primarily affected by internal air flow in ground vehicles?
3. Name and describe the three main regions in a simple pipe flow.

Application Questions

1. Why is the terminal velocity for turbulent flow reached faster than for laminar flow?

2. Consider a modification of Fig. 12.8 such that there is an overlap of the round ($p > p_{l\infty}$) region and the fillet ($p < p_{l\infty}$) region. Discuss the effect of this overlap on the internal flow and on the life of the pipe.

Exercise

Given that the air pressure is reduced by 10% after the front grille, find the pressure drop across the radiator of a car on the highway at a speed of 110 km/h

a. in a calm wind condition

b. in a headwind of 20 km/h

State your assumptions.

Projects

1. For a concept car design with a given fascia architecture, design the underhood air flow path for four scenarios:

    a. air exit at the vehicle rear

    b. air exit at a longitudinal location around the B-pillar

    c. air exit at the sides of the underbody

    d. air exit at the hood just before the windshield

    Compare the values of the Δp (the local static pressure—the exit pressure). Discuss the effect of each design on vehicle drag.

2. Consider the front and under the engine compartment of a car or truck, and perform the following tasks:

    a. identify stagnation points or regions

    b. describe the air flow into the grille

    c. determine whether the underhood air will exhaust effectively

    d. Identify the stagnation regions

References

1. Wong, J. Y., *Theory of Ground Vehicles*, 2nd. ed., John Wiley and Sons, Inc., 1993.
2. Sovran, G., Morel, T., and Mason, W. T. (eds.), *Aerodynamic Drag Mechanisms of Bluff Bodies and Road Vehicles*, Plenum Press, 1978.
3. Hoerner, S. F., *Fluid Dynamic Drag*, Chapter 9, Hoerner Published, 1965.
4. Emmelmann, H. J., "Aerodynamic Development and Conflicting Goals of Sub-Compacts—Outlined on the Opel Corsa," International Symposium Vehicle Aerodynamics, Wolfsburg, Germany, 1982.
5. Olson, M. E., and Schaub, U. W., "Aerodynamics of Trucks in Wind Tunnels: The Importance of Replicating Model Form, Model Detail, Cooling System and Test Conditions, Vehicle Aerodynamics, Wake Flows, Computational Fluid Dynamics, and Aerodynamic Testing," SAE SP-908, Paper 920345, 1992.
6. Schlichting, H., *Boundary Layer Theory*, McGraw-Hill, 1968.
7. Renn, V., and Gilhaus, A., "Aerodynamics of Vehicle Cooling Systems," *Journal of Industrial Engineering and Industrial Aerodynamics*, Vol. 22, 1986, pp. 339–346.
8. Buchheim, R., Deutenbach, K. R. and Luckoff, H. J., "Necessity and Premises for Reducing the Aerodynamic Drag of Future Passenger Cars," SAE Paper 810185, 1981.
9. Garrone, A., and Masoero, M., "Car Underside, Upper Body and Engine Cooling System Interactions, and Their Contributions to the Aerodynamic Drag," SAE Paper 860212, 1986.
10. Barnard, R. H., "Theoretical and Experimental Investigation of the Aerodynamic Drag Due to Automotive Cooling Systems," *Journal of Automobile Engineering*, Proceedings of the Institute of Mechanical Engineers (Part D), Vol. 24, D8, 2000, pp. 919–927.
11. Kargilis, A., *Design and Development of Automotive Engine Cooling System*, AlKar Engineering Company, 2001.

Chapter 13

Concept Ground Vehicles

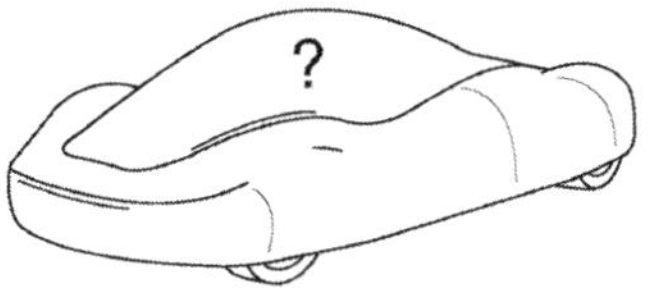

Vehicle design for improved aerodynamics is both a science and an art. We are faced with the dilemma of wanting conflicting phenomena—a high-speed laminar flow for drag reduction and a turbulent flow boundary layer to defer flow separation. The result is a compromise shaped by the specific mission of the vehicle. When the best design methods and styles for each of the three sections of a car (two sections in a truck) are optimally combined, we can theorize a concept vehicle—car, truck, bus. Naturally, such unified design features will need to be tested in soft design, in the wind tunnel, and sometimes on the intended road conditions for validity in application and for fine-tuning.

One of the ways to improve the performance of ground vehicles, at least in certain circumstances, is by automating the vehicle response to a prevailing situation. This approach, when properly implemented, eliminates such factors as the reaction time of the vehicle operator, the inherent human limitation in the instance being considered, and the skilled manpower that would be required for a manual response to the same situation. For example, accurate steering control improves the safe operation of vehicles, is more likely to accommodate an expansion of the speed range, allows for a narrower lane (only slightly wider than the vehicles), and is less accident prone than would be the case in manual control, since sensors and control devices generally have a much faster response time than humans and are not vulnerable to human fatigue. Also, the incorporation of continuous fault diagnostics makes the control instruments self-corrective, a feature that can contribute to the significant reduction of vehicle aerodynamic drag [13-1].

13.1 Passenger Car

Many studies have been done on the design of passenger cars, but there are still opportunities to improve the passenger car aerodynamic design. Drag reduction and vehicle stability are the chief aims of aerodynamic improvement. Depending on the vehicle's primary use, such improvements are best made by considering the car as a composite of three modules: the front section, the mid-section, and the rear section. The boundaries of these sections are then fused to give a seamless whole.

Mission specification: Passenger cars are primarily used for family outings, for transportation to work, and for commercial services such as taxis and limousines. The need for a *quick* drag reduction design in passenger cars is consistent with the short bursts of city traffic. In a streamlined body, the pressure drag is negligible at high speed (Reynolds number), and in general the viscous drag is minimal at low speed. Therefore, a streamlined car design gets the benefit of drag reduction both at high and low speeds (recall Table 1.1). It is important to stress streamline design over aesthetics, especially for taxis and cars that serve an exclusively functional purpose of transportation. Luxury vehicles are likely to incorporate design features that add to the aerodynamic drag and ultimately increase fuel consumption.

In Fig. 13.1, a concept car for the functional purpose of transportation is presented. It is tri-wheeled, with the front section starting off as an airplane cowling, fairing into an ellipsoid. The major axis of the ellipse from the top view coincides with the vehicle's longitudinal axis. The underbody is cut off parallel to the longitudinal axis, normal to the vertical axis, and at a distance z_1 from the bottom vertex such that $(z_1/z) \leq 1/4$, where z is the length of the minor axis of the ellipse from the front view (Fig. 13.2).

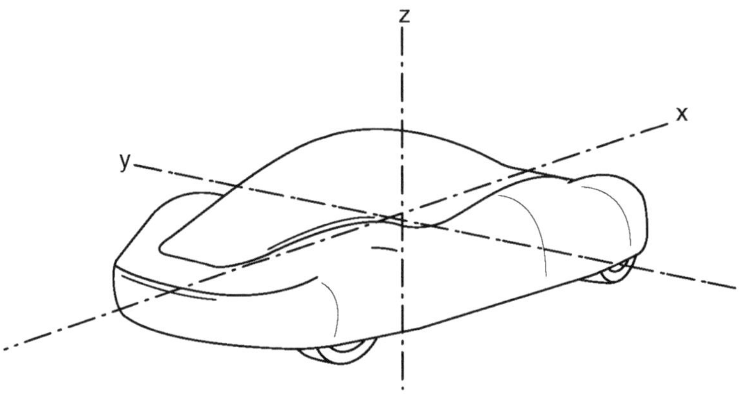

Fig. 13.1 Concept car

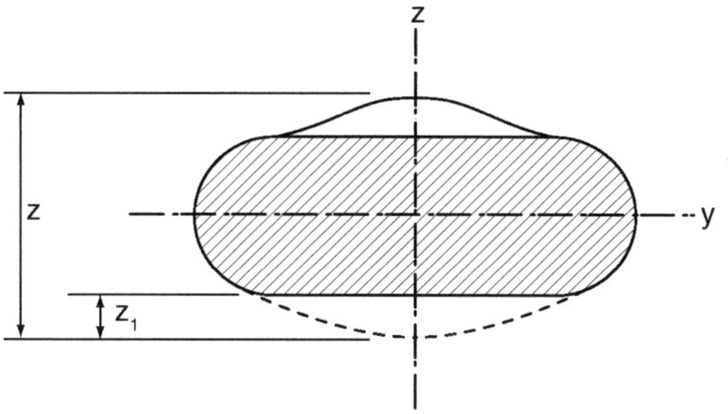

Fig. 13.2 Aft-driver lateral section through concept car

13.2 Race Car

A race car is typically smaller than a passenger car and is closer to the ground for stability during maneuvers. Also, the height-to-width ratio is less than that of a passenger car (Fig. 13.3). These features are in response to the mission specification of a race car.

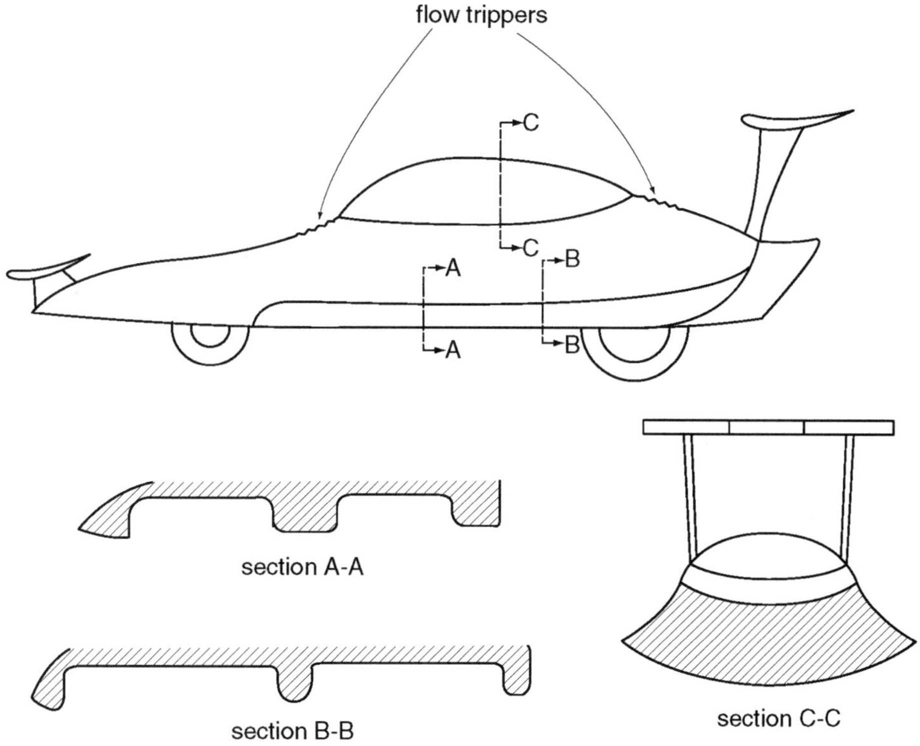

Fig. 13.3 Race car versus passenger car

The specific mission of a race car is to win races, that is, to be fast and as neutrally stable as possible in dynamic conditions. Therefore a race car needs to be fast and safely light, readily responding to rapid maneuvers. Greater attention is paid to stability than to drag reduction in race cars. This is understandable, since maneuverability and not fuel economy is the concern in race car design.

A balance needs to be struck between the light-weight design and stability. For example, a race car should be nose heavy to enhance the driver's control. But that adds to the weight. A compromise is to aerodynamically shape the front piece so that it enhances the car's stability, but it also assists in drag reduction; hence, the front spoiler.

Figure 13.4 is a concept race car. It includes both front and rear spoilers: the rear spoiler to give stability and the front spoiler to counter the nose-up tendency that may be caused by the rear spoiler's compelled positioning. The cockpit fairs seamlessly into the chassis, not only in the traditional longitudinal direction but also in the lateral direction, thus diffusing the tendency for a potential air flow separation soon after the cockpit. Also, flow trippers are installed at critical points where flow separation is imminent.

Underbody channels widen toward the rear, reducing the air pressure to local atmospheric condition at the car exit. The channel corners are also rounded to avoid the build-up of artificial walls in the flow path. Such a wall would add to both the vehicle drag and lift.

A sports car is, in essence, a modified race car. The modifications are typically in the form of power and torque reduction, and the elimination of some stability-enhancing features such as the front spoiler. Ironically, the stability-enhancing devices are add-ons to the basic car design.

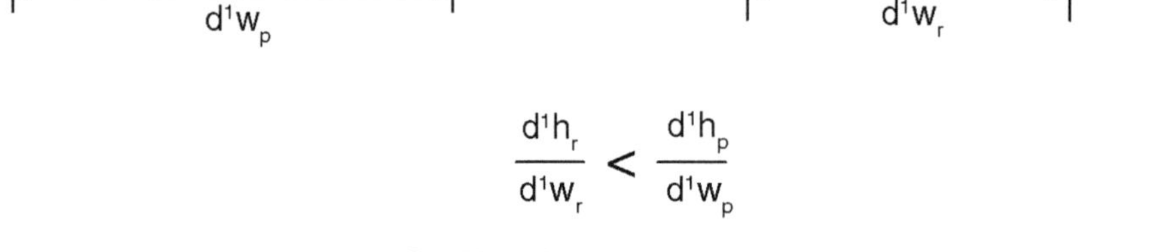

$$\frac{d^1h_r}{d^1w_r} < \frac{d^1h_p}{d^1w_p}$$

Fig. 13.4 Concept race car

13.3 Trucks and Trailers

There are ongoing studies on ways to improve the aerodynamic performance of trucks and trailers. Greater attention has been placed on the front section of these vehicles than on the aft section. This is understandable since up to 80% of the vehicle drag is experienced in the first third of the vehicle length. There are improvements that can be made to the rectangular-box aft sections. Flow separation regions, especially in full-length trailers, can be modified to delay separation without compromising the truck or trailer's carrying capacity. For example, laminar flow can be sustained in a truck or trailer by introducing a suction-type vertical depression on the sides of the bus or trailer, at a laminar distance from the leading edge of the flow. A repetition of this along the sides of the vehicle creates a continuous laminar flow even at high speeds.

Side skirts are gaining greater acceptance. They are placed on the underside of trailers, extending between the front and rear axles on each side of the trailer. For aerodynamic gain, the side skirts should narrow toward the rear rather than the current common design in which the enclosed width increases toward the rear.

A morphing of the trailer rear into a modified hemispherical enclosure helps to reduce separation at the rear. This has the dual benefits of drag reduction at the rear of the truck and reduced vortex generation.

13.4 Motorcycles

The average length of an entire motorcycle is about 67 in. (1700 mm). This is about the length of the front section of a medium car. Therefore the aerodynamic focus is primarily equivalent to the front section design consideration for a car, with less attention paid to the profiling of the front area—a motorcycle's frontal area is about one-fifth the frontal area of a medium car.

Mission specification: Aside from the excellent gas mileage (not the specific gas mileage) when compared with a car, maneuverability is the primary interest in a motorcycle design. There is a strong interaction between drag and maneuverability in a motorcycle because of the rider's exposure to the elements. Even at lower speeds the drag created by a motorcycle and its rider has a substantial effect on performance and stability. At higher speeds, drag is the most prominent parameter in determining a motorcycle's overall performance.

Modern motorcycles have incorporated a number of drag reduction and stability-improving devices. The windshield is one such device. The proper windshield for a motorcycle depends on the intended use of the motorcycle. For example, a short windshield, as used in sport motorcycles, will only increase the drag and create discomfort for a rider that attempts to ride it as a tour or standard motorcycle. The guideline is that the height of the windshield should be at least the same as that of the top of the helmet of the rider. When there is no windshield, the handlebar should be as shown in Fig. 13.5. This is in contrast to the traditional bluff-body handlebar that, in spite of the aerodynamic front shell, inevitably increases the drag.

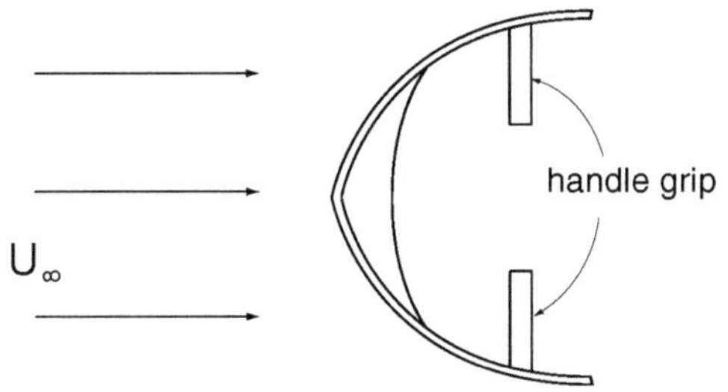

Fig. 13.5 Motorcycle handlebar concept design

Front fenders are also flow diverters. The fenders should steer slightly backward and away from the longitudinal axis.

13.5 Train

High-speed trains have incorporated advanced aerodynamic features that have translated to higher speed. But the high speeds bring with them other challenges, including compressibility and noise. For passenger comfort, one of the solutions to the increasing noise may be the introduction of better noise attenuators in the coach inner wall and in the upholstery.

Passenger trains utilized for local commuting generally have a blunt engine module. This should be improved to a shape approaching that of the *bullet train*. Just this modification of the leading car may suffice for the aerodynamic improvement of local commute trains.

13.6 Severe Service and Off-Road Vehicles

Most of the aerodynamic improvement that can be made to severe service vehicles is in the front section. Aerodynamic design of the leading edge, a windshield that slants backward as it rises from the hood-windshield interface to the windshield-roof edge of the vehicle, and improved underflow at the vehicle leading edge all contribute to drag reduction in any vehicle, including severe service vehicles.

Off-road vehicles are usually considered as one small piece (such as an in-plant cart) or one massive piece (as in an armored tank). In either case, most of the aerodynamic improvement that can be made is in the front portion of the vehicle.

13.7 Aerocars and Amphibians

A concept vehicle that is likely to gain market share soon is the aerocar. Also called the roadable airplane or flying car, the aerocar is typically a two-seater that, in principle, transforms to an airplane and rescues the occupants from the ever-increasing traffic congestion in the cities and on the highways. Aerodynamic principles are already incorporated into the blueprint of the aerocar, so it is unlikely that the aerocar will go through the long and slow aerodynamic phases that road vehicles have traveled to reach a finer aerodynamic design.

Because of its low cost compared to a helicopter, the aerocar may soon be a common sight in spite of the disadvantages of increased traffic (and hence potential for accidents) in the skies and expanded air safety regulations.

Like aerocars, amphibians have an appeal as vehicles to expedite transportation of goods and passengers. Most of the same aerodynamic considerations for passenger cars would apply to amphibians, but with a greater emphasis on the underbody design. Also, the amphibian is subjected to a two-phase flow environment when on water. The significance of this is in the determination of the total drag (above and below the waterline) and the downforce. Amphibians are more likely to be corporate or public transport vehicles than privately owned.

Reference

1. Shladover, S., "Automated Driving of Trucks and Buses: Opportunities for Increasing Productivity and Safety while Reducing Fuel Use and Emissions," *The Aerodynamics of Heavy Vehicles: Trucks, Buses and Trains*, Monterey, California, December 2–6, 2002, p. 563.

Nomenclature

D	drag
L	lift
T	thrust
W	weight, normal load
R	resistive force
M	moment
m	mass
μ	coefficient of rolling resistance
g	acceleration due to gravity
l	length
A, S	area
P	engine power
p	pressure
ρ	density
v	speed
α	angle of attack

Conversion Table

Quantity	Symbol	SI to U.S. Conversion	U.S. to SI Conversion
Length	*l*, *L*	1 m = 3.2808 ft	1 ft = 0.3048 m
Area	*A*, *S*	1 m^2 = 10.7639 ft^2	1 ft^2 = 0.092903 m^2
Velocity	v	1 m/s = 3.2808 ft/s	1 ft/s = 0.3048 m/s
Acceleration	a	1 m/s^2 = 3.2808 ft/s^2	1 ft/s^2 = 0.3048 m/s^2
Force	*F*	1 N = 0.2248 lb_f	1 lb_f = 4.4482 N
Mass	*m*	1 kg = 2.20462 lb_m	1 lb_m = 0.4359 kg
Density	*ρ*	1 kg/m^3 = 0.06243 lb_m/ft^3	1 lb_m/ft^3 = 16.018 kg/m^3
Pressure	*p*	1 N/m^2 = 1.45038×10^{-4} lb_f/in^2	1 lb_f/in^2 = 6894.76 N/m^2
Energy	*E*	1 J = 0.7376 ft-lb_f	1 ft-lb_f = 1.3558 J
Power	*P*	1W = 1.341022×10^{-3} hp = 0.7376 ft-lb_f/s	1 hp (= 550 ft-lb_f/s) = 746 W
Moment of a force	*M*	1 N-m = 0.7376 lb_f-ft	1 lb_f-ft =1.3558 N-m

Index

About the Author

Dr. T. Yomi Obidi is currently the director of Research and Development at Tamayi Kristi, LLC. His work experience, both in academics and in industry, spans over 35 years. Dr. Obidi's academic career includes faculty positions at Ryerson Polytechnic Institute, State University College at Buffalo, University of Ibadan, and Embry-Riddle Aeronautical University, where he was also the director of the Wind Tunnel Laboratory. He was a cooling system engineer with primary responsibilities for the inline (I6) engines at International Truck and Engine Corporation, and a cooling system product engineer in the Small Car Platform at DaimlerChrysler. He worked as an aerodynamic assistant consultant on Piper aircraft through the University of Florida. He has served as an adjunct faculty member in the Mathematics Department at Roosevelt University, and he is currently on the adjunct faculty of the City Colleges of Chicago, where he teaches Mathematics and Physics.

Dr. Obidi received a B.S. in Aeronautical Engineering from Embry-Riddle Aeronautical University, an M.S. in Aerospace Engineering from University of Florida, and a Ph.D. in Mechanical Engineering from the State University of New York. He is licensed as a Professional Engineer in the state of Illinois and in the province of Ontario, Canada. He is a member of the Society of Automotive Engineers.